Holt
Mathematics

Chapter 6 Resource Book

HOLT, RINEHART AND WINSTON

A Harcourt Education Company

Orlando • **Austin** • New York • San Diego • London

ISBN 0-03-078302-X

4 5   170   09 08 07

CONTENTS

Holt Mathematics

Date ____________

Dear Family,

In this chapter, your child will learn how to work with concepts related to percents, including fractions, decimals, finding a percent of a number, solving percent problems, and computing simple interest. The concepts of fractions and percents are applied to computers and the internet, nutrition, discounts in stores, and sports statistics—topics of interest to many middle school students.

A **percent** is a comparison of a number to 100. For example, 40% can be written as the ratio 40 to 100, or $\frac{40}{100}$. Percents can be written as fractions or decimals.

Write the percent as a fraction in simplest form.

$35\% = \frac{35}{100}$ *Write the percent as a fraction with a denominator of 100.*

$\frac{35}{100} = \frac{7}{20}$ *Write the fraction in simplest form.*

Write the **percent as a decimal.**

$43\% = \frac{43}{100} = 0.43$

Here are some common fraction and percent pairs that your child will learn.

Percent	10%	20%	25%	$33\frac{1}{3}\%$	50%
Fraction	$\frac{1}{10}$	$\frac{1}{5}$	$\frac{1}{4}$	$\frac{1}{3}$	$\frac{1}{2}$

You can set up and solve a proportion to find the **percent of a number.**

$$\frac{\text{part}}{\text{whole}} \longrightarrow \frac{n}{100}$$

Find the percent of the number.

67% of 90

$\frac{67}{100} = \frac{n}{90}$ *Write a proportion.*

$67 \cdot 90 = 100 \cdot n$ *The cross products are equal.*

$6{,}030 = 100n$ *Simplify.*

$\frac{6{,}030}{100} = \frac{100n}{100}$ *Divide each side by 100 to isolate the variable.*

$60.3 = n$

Holt Mathematics

Your child will also learn to solve percent problems and see applications in areas such as health, music, and history.

45 is what percent of 90?

$$\frac{n}{100} = \frac{45}{90}$$ *Write a proportion.*

$n \bullet 90 = 100 \bullet 45$ *The cross products are equal.*

$90n = 4{,}500$ *Simplify.*

$$\frac{90n}{90} = \frac{4{,}500}{90}$$ *Divide each side by 90 to isolate the variable.*

$n = 50$

45 is 50% of 90.

You can find the **percent of change** by using the following formula.

$$\text{percent of change} = \frac{\text{amount of change}}{\text{original amount}}$$

Find the percent of change when 25 is decreased to 20.

$25 - 20 = 5$ *Find the amount of change.*

$\text{percent of change} = \frac{5}{25}$ *Substitute values into formula.*

$= 0.20$ *Divide.*

$= 20\%$ *Write the decimal as a percent.*

The percent decrease is 20%.

Simple interest is money paid on only the principal amount, which is the amount of money deposited or loaned. To solve problems involving simple interest, you use this formula:

$I = P \cdot r \cdot t$, where $I =$ interest, $p =$ principal, $r =$ rate of interest per year as a decimal, and $t =$ time in years.

Find the missing value.

$I = \$600$, $P = \$5{,}000$, $r = \blacksquare$, $t = 3$

$$I = P \cdot r \cdot t$$

$600 = 5{,}000 \cdot r \cdot 3$ *Substitute.*

$600 = 15{,}000 \cdot r$ *Multiply.*

$$\frac{600}{15{,}000} = \frac{15{,}000}{15{,}000} \cdot r$$ *Divide by 15,000 to isolate the variable.*

$0.04 = r$

The interest rate is 4%.

For additional resources, visit go.hrw.com and enter the keyword MS7 Parent.

Holt Mathematics

LESSON 6-1 **Practice A**
Percents

Write the fraction of the grid that is shaded. Then write the percent.

1. 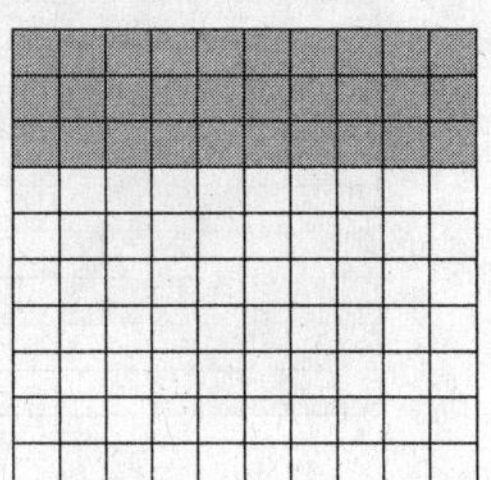

$$\frac{30}{100} = \quad \%$$

2. 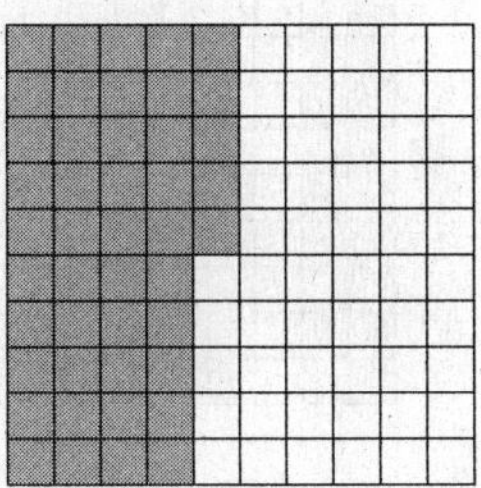

$$\frac{\quad}{100} = \quad \%$$

3. 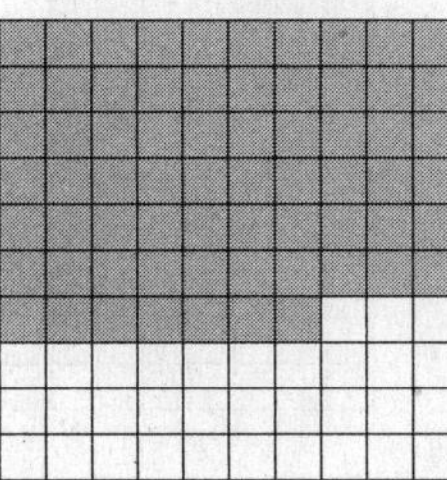

$$\frac{\quad}{100} = \quad \%$$

Write each percent as a fraction in simplest form.

4. 20% **5.** 32% **6.** 90% **7.** 43%

_______________ _______________ _______________ _______________

8. 48% **9.** 58% **10.** 71% **11.** 80%

_______________ _______________ _______________ _______________

12. 28% **13.** 85% **14.** 42% **15.** 52%

_______________ _______________ _______________ _______________

Write each percent as a decimal.

16. 19% **17.** 43% **18.** 7% **19.** 62%

_______________ _______________ _______________ _______________

20. 23% **21.** 36% **22.** 59% **23.** 3%

_______________ _______________ _______________ _______________

24. 72.5% **25.** 9.8% **26.** 7.04% **27.** 12.49%

_______________ _______________ _______________ _______________

Holt Mathematics

Practice B
Percents

Write the percent modeled by each grid.

1. 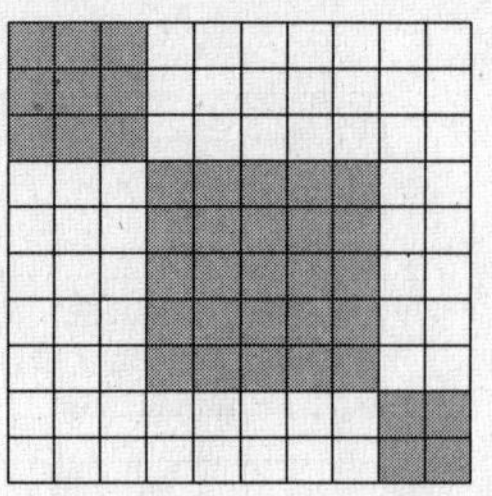

2. 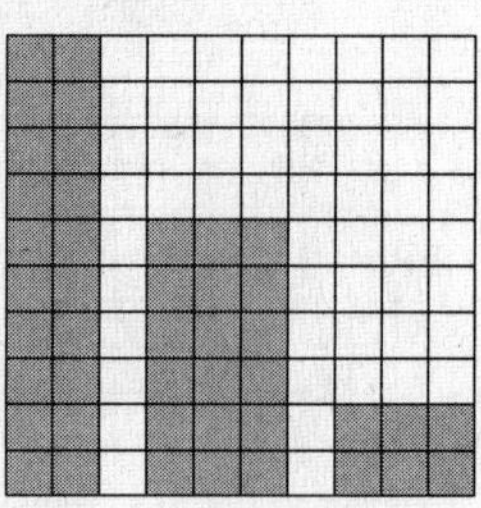

3. 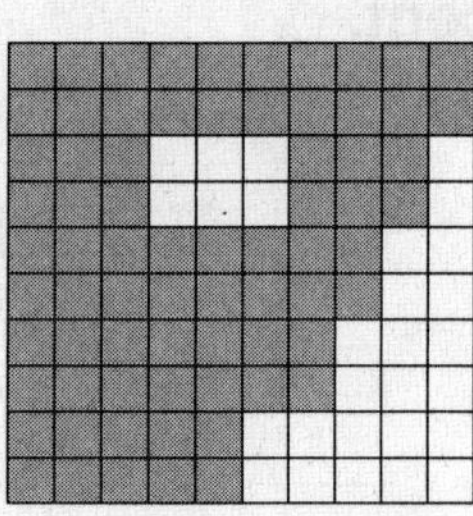

____________   ____________   ____________

Write each percent as a fraction in simplest form.

4. 16%   5. 49%   6. 20%   7. 15%

____________   ____________   ____________   ____________

8. 18%   9. 60%   10. 35%   11. 46%

____________   ____________   ____________   ____________

12. 86%   13. 79%   14. 56%   15. 45%

____________   ____________   ____________   ____________

Write each percent as a decimal.

16. 33%   17. 57%   18. 46%   19. 6%

____________   ____________   ____________   ____________

20. 4.7%   21. 13.2%   22. 75.8%   23. 4%

____________   ____________   ____________   ____________

24. 1.16%   25. 27.05%   26. 93.01%   27. 7.9%

____________   ____________   ____________   ____________

Holt Mathematics

LESSON 6-1

Practice C
Percents

Write the percent modeled by each grid.

1. 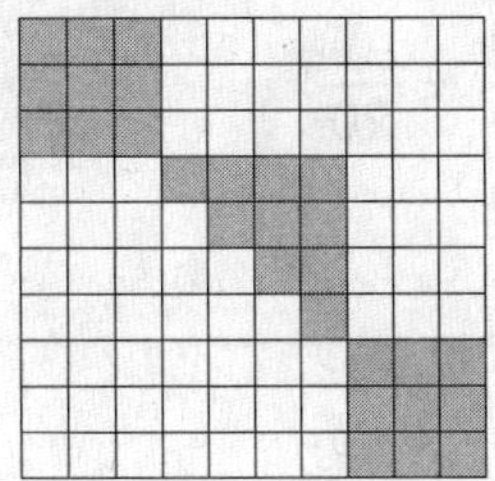

2. 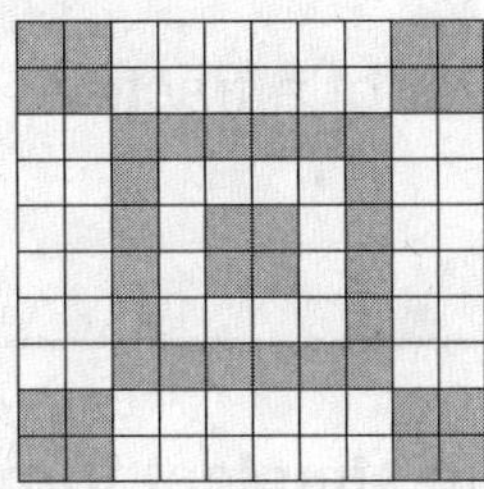

3. 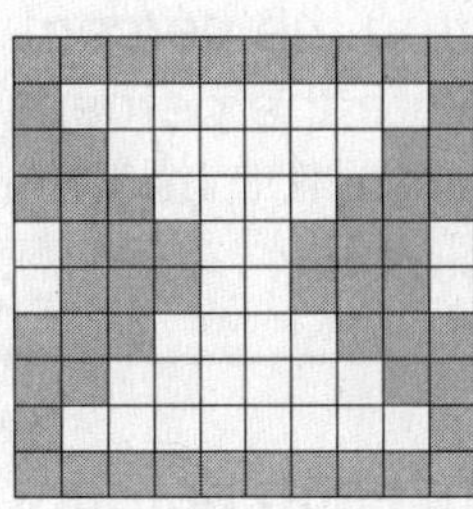

_______________ _______________ _______________

Write each percent as a fraction in simplest form and as a decimal.

4. 29%

5. 87.5%

6. 7.5%

7. 64%

_______________ _______________ _______________ _______________

8. 62.5%

9. 34%

10. 22.5%

11. 12.5%

_______________ _______________ _______________ _______________

12. 3.6%

13. 5.9%

14. 7.2%

15. 20.5%

_______________ _______________ _______________ _______________

Compare. Write <, >, or =.

16. $\dfrac{19}{100}$ ▨ 9%

17. $\dfrac{44}{100}$ ▨ 44%

18. $\dfrac{38}{100}$ ▨ 39%

_______________ _______________ _______________

19. $\dfrac{12}{25}$ ▨ 40%

20. $\dfrac{19}{20}$ ▨ 96%

21. $\dfrac{27}{50}$ ▨ 52%

_______________ _______________ _______________

22. $\dfrac{7}{40}$ ▨ 17.5%

23. $\dfrac{5}{12}$ ▨ 41%

24. $\dfrac{1}{16}$ ▨ 6.5%

_______________ _______________ _______________

Holt Mathematics

LESSON 6-1 **Reteach**
Percents

To change a percent to a fraction:
- drop the percent symbol;
- write the percent as the numerator of a fraction; $45\% = \dfrac{45}{100} = \dfrac{9}{20}$
- write 100 as the denominator;
- simplify.

Write each percent as a fraction in simplest form.

1. $38\% = \dfrac{}{100} =$ _______

2. $20\% = \dfrac{}{100} =$ _______

3. $70\% = \dfrac{}{100} =$ _______

4. $16\% = \dfrac{}{100} =$ _______

5. $36\% = \dfrac{}{100} =$ _______

6. $8\% = \dfrac{}{100} =$ _______

7. $15\% =$ _______

8. $53\% =$ _______

9. $24\% =$ _______

10. $17\% =$ _______

To change a percent to a decimal:
- drop the percent symbol; $45\% = .45. = 0.45$
- move the decimal point two places to the left. $7\% = .07. = 0.07$

Write each percent as a decimal.

11. 58%   **12.** 93%   **13.** 15%   **14.** 9%

_______   _______   _______   _______

15. 26%   **16.** 2%   **17.** 80%   **18.** 1%

_______   _______   _______   _______

19. 23.5%   **20.** 9.6%   **21.** 40.7%   **22.** 7.03%

_______   _______   _______   _______

Holt Mathematics

LESSON 6-1 Challenge
Complementary Events

The probability of an event can be expressed as a percent, a fraction, or a decimal.

When two events are the only two events that can occur, they are called **complementary events**. If two events are complementary, the sum of their probabilities is one. For example, suppose the probability of rain is 60%, or 0.6: $P(rain) = 0.6$. That means the probability of it not raining is 40%, or 0.4: P(not rain) = 0.4. So, $P(rain\ or\ not\ rain) = 0.6 + 0.4 = 1$.

Draw a line to connect the probability for each Event A to the probability for an Event B that could be a complementary event.

Event A	Event B	
$P(A) = 45\%$	$P(B) = 0.29$	F
	$P(B) = \frac{5}{8}$	T
$P(A) = 70\%$	$P(B) = 0.35$	L
	$P(B) = 0.175$	N
$P(A) = 42\%$	$P(B) = \frac{13}{25}$	E
	$P(B) = \frac{11}{20}$	A
$P(A) = 17.5\%$	$P(B) = \frac{12}{25}$	I
	$P(B) = 0.42$	S
$P(A) = 65\%$	$P(B) = \frac{3}{10}$	R
	$P(B) = \frac{9}{20}$	H
$P(A) = 37.5\%$	$P(B) = \frac{33}{40}$	U
$P(A) = 48\%$	$P(B) = \frac{29}{50}$	M

Look for the probabilities that do <u>not</u> have lines drawn to them. Use the letters next to those probabilities to write the answer to the riddle. You may use each letter as many times as you wish.

What language do sharks speak?

____ ____ ____ ____ ____ ____ ____ ____

Holt Mathematics

LESSON 6-1 Problem Solving
Percents

Write the correct answer.

1. In 2003, 68% of the T.V. set owners in the United States had cable television. Write this percent as a fraction in simplest form and as a decimal.

2. In 2004, 27% of Internet users were in the United States. What percent of Internet users were in countries other than the United States?

3. In a survey, 46% of men said they spend fewer than 5 hours shopping for gifts for the holidays. Write this percent as a fraction in simplest form and as a decimal.

4. In a survey, 59% of a group of people aged 18-29 said that they do not have enough time to do what they want. What percent of those surveyed feel that they do have enough time do what they want? Write your answer as a percent and as a decimal.

Choose the letter for the best answer.

The table shows the percent of adults who participated in selected leisure activities two or more times per week.

5. Express the percent of adults who dined out two or more times per week as a fraction.

A $\frac{1}{100}$ C $\frac{1}{10}$

B $\frac{1}{50}$ D $\frac{1}{5}$

6. Express the percent of adults who did crossword puzzles two or more times per week as a decimal.

F 0.07   H 0.5

G 0.05   J 0.7

7. What fraction of adults played video games fewer than two times per week?

A $\frac{1}{20}$ C $\frac{19}{50}$

B $\frac{1}{50}$ D $\frac{19}{20}$

Adult Participation in Selected Leisure Activities in 2003

Activity	Percent
Crossword puzzles	7%
Dining out	10%
Reading books	21%
Surfing the net	18%
Video games	5%

8. Which decimal represents the percent of adults who surfed the net fewer than two times per week?

F 0.018   H 0.18

G 0.05   J 0.82

Holt Mathematics

Reading Strategies
Multiple Representations

Percent means "per hundred." You can use a hundred grid to picture percents.

The shaded part of the grid
is 20% → Read: "20 percent."

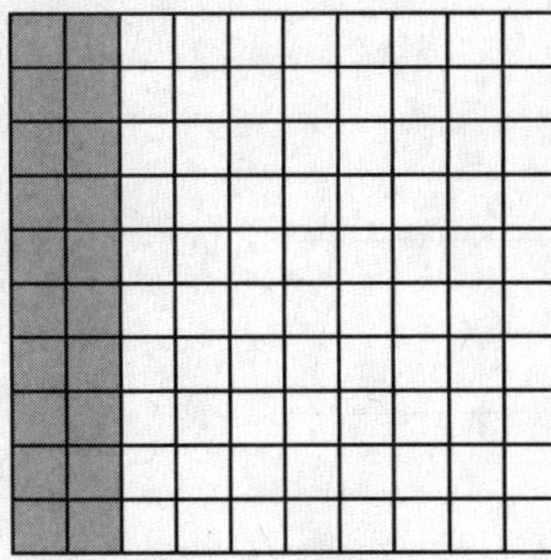

Percents can be written as fractions and as decimals.

20% means "20 per hundred" or $\frac{20}{100}$. The fraction can be written
in simplest form. → $\frac{1}{5}$

$\frac{20}{100}$ is read → "20 hundredths" and can be written → 0.20.

The shaded part of the hundred grid can be represented in these
ways:

20% → $\frac{20}{100}$ → $\frac{1}{5}$ → 20 hundredths → 0.20

Answer the following questions.

1. What does the word "percent" mean? _________________________________

2. What two other ways can you write percents? _________________________

3. Shade 40% of the hundred grid.

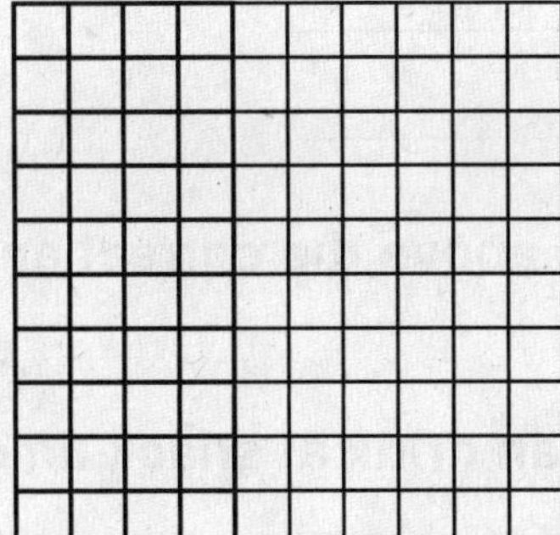

4. Write the shaded part as a decimal. _________________________________

5. Write the shaded part as a fraction. _______________ or _______________

6. If $\frac{1}{2}$ of the grid were shaded, what percent would that be? _______________

Holt Mathematics

Puzzles, Twisters, & Teasers

LESSON 6-1

Now You See It...

Write each percent as a fraction in simplest form.

1. 65% O _______________

2. 20% E _______________

3. 72% T _______________

4. 63% R _______________

5. 6% V _______________

6. 24% K _______________

Write each percent as a decimal.

7. 3.62% L _______________

8. 60% A _______________

9. 52% I _______________

10. 6.3% M _______________

11. 45% P _______________

12. 36.2% D _______________

Write the letter on the line above the correct answer to solve the riddle.

What does the Invisible Man drink at snack times?

He drinks ____ ____ ____ ____ ____ ____ ____ ____ ____ ____

$\frac{1}{5}$ $\frac{3}{50}$ 0.6 0.45 $\frac{13}{20}$ $\frac{63}{100}$ 0.6 $\frac{18}{25}$ $\frac{1}{5}$ 0.362

____ ____ ____ ____

0.063 0.52 0.0362 $\frac{6}{25}$

10

Holt Mathematics

Practice A
Fractions, Decimals, and Percents

Complete.

1. $0.16 = \dfrac{}{100} = \%$

2. $0.09 = \dfrac{}{100} = \%$

3. $\dfrac{3}{5} = \dfrac{}{100} = \%$

4. $\dfrac{21}{25} = \dfrac{}{100} = \%$

Write each decimal as a percent.

5. 0.34

6. 0.08

7. 0.19

8. 0.64

9. 0.561

10. 0.049

11. 0.105

12. 0.9

Write each fraction as a percent.

13. $\dfrac{1}{4}$

14. $\dfrac{7}{50}$

15. $\dfrac{15}{16}$

16. $\dfrac{9}{20}$

17. $\dfrac{7}{8}$

18. $\dfrac{9}{40}$

19. $\dfrac{11}{50}$

20. $\dfrac{4}{5}$

Decide whether to use pencil and paper, mental math, or a calculator. Then solve.

21. In a survey, 25 students were asked whether they prefer orange juice or grapefruit juice. Seventeen students said they prefer orange juice. What percent of the students surveyed said they prefer orange juice?

____

____

Holt Mathematics

LESSON 6-2 — Practice B
Fractions, Decimals, and Percents

Write each decimal as a percent.

1. 0.17   **2.** 0.56   **3.** 0.04   **4.** 0.7

_______   _______   _______   _______

5. 0.025   **6.** 0.803   **7.** 0.3   **8.** 0.072

_______   _______   _______   _______

Write each fraction as a percent.

9. $\frac{13}{40}$ **10.** $\frac{3}{5}$ **11.** $\frac{3}{20}$ **12.** $\frac{5}{12}$

_______   _______   _______   _______

13. $\frac{5}{16}$ **14.** $\frac{3}{80}$ **15.** $\frac{5}{6}$ **16.** $\frac{19}{25}$

_______   _______   _______   _______

Decide whether pencil and paper, mental math, or a calculator is most useful when solving the following problems. Then solve.

17. In a survey, 60 baseball fans were asked whether they thought the designated hitter rule should be changed. Forty-one fans thought the rule should be changed. What percent of the fans surveyed said that the designated hitter rule should be changed?

____

____

18. The police use a speed gun to monitor one part of a highway. During one hour, 6 out of 25 cars were traveling above the speed limit. What percent of the cars were traveling above the speed limit?

____

____

Holt Mathematics

LESSON 6-2 Practice C
Fractions, Decimals, and Percents

Write each decimal as a percent.

1. 0.04   2. 0.975   3. 0.031   4. 0.409

______   ______   ______   ______

5. 0.378   6. 0.4   7. 0.524   8. 0.067

______   ______   ______   ______

Write each fraction as a percent.

9. $\frac{13}{80}$ 10. $\frac{7}{9}$ 11. $\frac{17}{20}$ 12. $\frac{5}{8}$

______   ______   ______   ______

13. $\frac{39}{40}$ 14. $\frac{4}{11}$ 15. $\frac{11}{25}$ 16. $\frac{9}{50}$

______   ______   ______   ______

Compare. Write <, >, or =.

17. 30% ▮ $\frac{1}{3}$ 18. $\frac{3}{5}$ ▮ 60%   19. $\frac{7}{16}$ ▮ 44%

______   ______   ______

20. 0.065 ▮ 65%   21. 0.507 ▮ 50%   22. $\frac{7}{40}$ ▮ 20%

______   ______   ______

Decide whether pencil and paper, mental math, or a calculator is most useful when solving the following problem. Then solve.

23. In a survey, 80 students were asked to name their favorite subject. Thirty students said that English was their favorite. What percent of the students surveyed said that English was their favorite subject?

____

____

13

Holt Mathematics

LESSON 6-2 — Reteach
Fractions, Decimals, and Percents

To change a decimal to a percent:
- move the decimal point two places to the right; $0.07 = .07. = 7\%$
- write the % symbol after the number.

Write each decimal as a percent.

1. 0.34

2. 0.06

3. 0.93

4. 0.57

5. 0.8

6. 0.734

7. 0.082

8. 0.225

9. 0.604

10. 0.09

11. 0.518

12. 0.039

To change a fraction to a percent:
- Find an equivalent fraction with a denominator of 100.
- Use the numerator of the equivalent fraction as the percent.

$$\frac{8}{25} = \frac{x}{100}$$
$$\frac{8 \cdot 4}{25 \cdot 4} = \frac{32}{100}$$
$$\frac{8}{25} = \frac{32}{100} = 32\%$$

Think: $100 \div 25 = 4$. So, multiply the numerator and denominator by 4.

Write each fraction as a percent.

13. $\frac{3}{10}$

14. $\frac{2}{50}$

15. $\frac{7}{20}$

16. $\frac{1}{5}$

17. $\frac{1}{8}$

18. $\frac{3}{25}$

19. $\frac{3}{4}$

20. $\frac{23}{40}$

21. $\frac{11}{20}$

22. $\frac{43}{50}$

23. $\frac{24}{25}$

24. $\frac{7}{8}$

Holt Mathematics

LESSON 6-2

Challenge
Climbing a Percent Pyramid

Begin by changing each ratio in the pyramid to a percent. Then climb to the top by comparing the percents as you go.

The percents must increase as you climb to the top. Each brick on your path must touch the previous brick. Circle the ratios in the path that leads to the top.

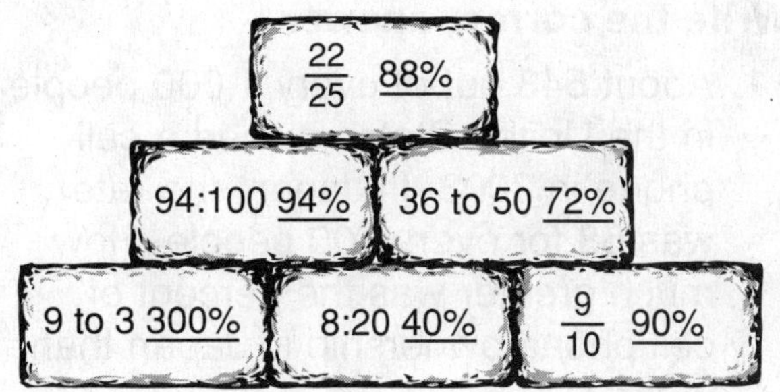

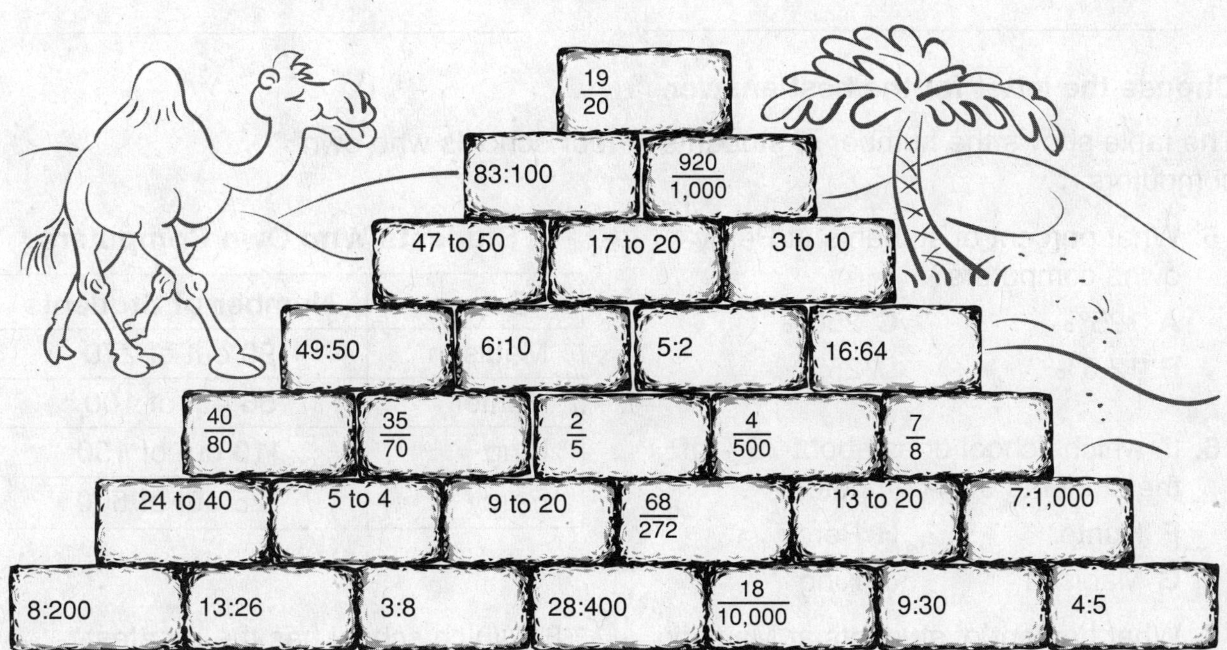

Holt Mathematics

Problem Solving
Fractions, Decimals, and Percents

Write the correct answer.

1. About 543 out of every 1,000 people in the United States owned a cell phone in 2003. In Japan, the rate was 68 for every 100 people. How much greater was the percent of cell phone ownership in Japan than in the U.S.?

2. In 2002, the adult population of the United States was about 206 million. About 113 million people participated in an exercise program. To the nearest percent, what percent of the adult population participated in an exercise program?

3. When asked about their favorite Thanksgiving leftover, $\frac{1}{20}$ of the people said vegetables and $\frac{7}{100}$ said mashed potatoes. Which food was more popular and by what percent?

4. In a survey, 80 people were asked whether they thought the speed limit for interstate highways should be raised. Twenty-five people said the speed limit should be raised. What percent of people did not think that the speed limit should be raised?

Choose the letter for the best answer.

The table shows the number of students in four schools who own computers.

5. What percent of students at Percy owns computers?

  A 125%   **C** 250%

  B 12.5%   **D** 25%

Students Who Own Computers

School	Number of Students
Madison	90 out of 270
Hunter	56 out of 100
King	110 out of 150
Percy	125 out of 500

6. In which school does about 73% of the students own computers?

  F Hunter   **H** Percy

  G Madison   **J** King

7. What percent of students at Madison do not own computers? Round to the nearest tenth of a percent.

  A 3.3%   **C** 33.3%

  B 9.9%   **D** 66.7% [d]

8. Which school has the greatest percent of students who own computers?

  F Hunter   **H** Percy

  G Madison   **J** King

Holt Mathematics

LESSON 6-2 **Reading Strategies**
Use a Diagram

You can use a diagram to understand the relationships among fractions, decimals, and percents.

Write 120 out of 400 as a fraction in tenths, as a decimal, and as a percent.

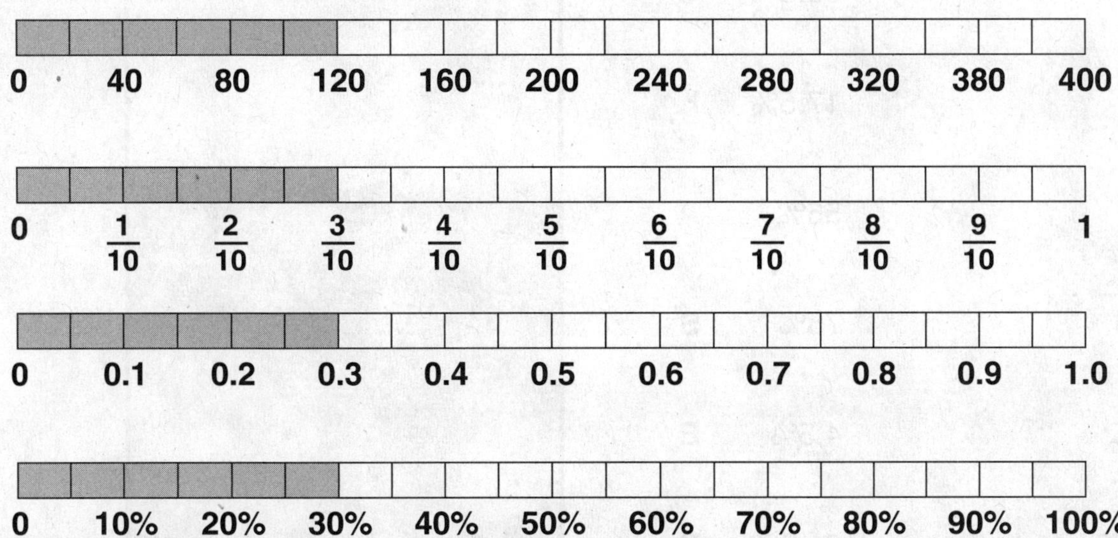

The diagram shows that 120 out of 400 can be written as $\frac{3}{10}$, 0.3, or 30%.

Use the diagram to answer each question.

1. What fraction in tenths can you write for 280 out of 400?

____

2. What percent can you write for 280 out 400?

____

3. What decimal can you write for 180 out of 400?

____

4. What percent can you write for 180 out 400?

____

5. How could you change or add to the diagram above so that you could use it to find the percent that is equal to 110 out of 200?

____

____

Holt Mathematics

Puzzles, Twisters, & Teasers
Pick a Percent

Draw a line from each fraction or decimal to the equivalent percent.

1. $\dfrac{3}{4}$ 55.$\overline{5}$%   **U**

2. $\dfrac{5}{9}$ 17.5%   **I**

3. $\dfrac{13}{16}$ 55%   **A**

4. 0.84   75%   **S**

5. 0.55   4.5%   **E**

6. 0.075   81.25%   **B**

7. $\dfrac{7}{40}$ 82.5%   **S**

8. $\dfrac{9}{20}$ 7.5%   **R**

9. 0.045   84%   **M**

10. 0.825   45%   **N**

Write the letter on the line above the matching problem number to solve the riddle.

What kind of sandwiches do sailors eat?

___ ___ ___ ___ ___ ___ ___ ___ ___ ___
1.   2.   3.   4.   5.   6.   7.   8.   9.   10.

Holt Mathematics

Practice A

LESSON 6-3

Estimate with Percents

Which fraction is the best estimate for the given percent? Choose the letter of the best answer.

1. 24%   A $\frac{1}{2}$ B $\frac{1}{3}$ C $\frac{1}{4}$ D $\frac{1}{5}$

2. 73%   F $\frac{4}{5}$ G $\frac{3}{4}$ H $\frac{2}{3}$ J $\frac{1}{2}$

3. 51%   A $\frac{1}{2}$ B $\frac{2}{3}$ C $\frac{3}{4}$ D $\frac{4}{5}$

4. 32%   F $\frac{1}{5}$ G $\frac{1}{4}$ H $\frac{1}{3}$ J $\frac{1}{2}$

5. 39%   A $\frac{1}{5}$ B $\frac{2}{5}$ C $\frac{3}{5}$ D $\frac{4}{5}$

6. 12%   F $\frac{1}{3}$ G $\frac{1}{4}$ H $\frac{1}{5}$ J $\frac{1}{10}$

7. 19%   A $\frac{1}{2}$ B $\frac{1}{3}$ C $\frac{1}{4}$ D $\frac{1}{5}$

Use a fraction to estimate the percent of each number.

8. 24% of 80   9. 73% of 44   10. 51% of 120   11. 32% of 90

_____________   _____________   _____________   _____________

12. 39% of 50   13. 12% of 40   14. 19% of 30   15. 58% of 200

_____________   _____________   _____________   _____________

Use 1% or 10% to estimate the percent of each number.

16. 39% of 25   17. 11% of 20   18. 3% of 200   19. 71% of 50

_____________   _____________   _____________   _____________

20. 9% of 40   21. 49% of 60   22. 4% of 50   23. 89% of 30

_____________   _____________   _____________   _____________

Holt Mathematics

Practice B
Estimate with Percents

Use a fraction to estimate the percent of each number.

1. 21% of 82   **2.** 35% of 42   **3.** 47% of 164   **4.** 9% of 68

_______________   _______________   _______________   _______________

5. 65% of 78   **6.** 11% of 92   **7.** 26% of 124   **8.** 89% of 51

_______________   _______________   _______________   _______________

9. 77% of 198   **10.** 5% of 75   **11.** 31% of 148   **12.** 53% of 539

_______________   _______________   _______________   _______________

13. In 2004, about $38 out of every $100 spent on advertising was spent on television advertising. The amount spent on radio advertising was about 21% as much as was spent on television advertising. How much of every $100 spent on advertising was spent on radio advertising? _______________

Use 1% or 10% to estimate the percent of each number.

14. 32% of 46   **15.** 81% of 36   **16.** 15% of 44   **17.** 21% of 62

_______________   _______________   _______________   _______________

18. 3% of 72   **19.** 62% of 88   **20.** 12% of 48   **21.** 65% of 124

_______________   _______________   _______________   _______________

22. 18% of 147   **23.** 5% of 837   **24.** 37% of 213   **25.** 2% of 188

_______________   _______________   _______________   _______________

26. The Fresh Acres Swim Club has a $35,000 budget for pool maintenance this year. The club members have agreed to raise the budget by 4%. Estimate the pool maintenance budget for next year. _______________

Holt Mathematics

Name _______________________________________ Date __________ Class __________

Practice C

Estimate with Percents

Use a fraction to estimate the percent of each number.

1. 23% of 53　　　**2.** 35% of 37　　　**3.** 46% of 113　　　**4.** 8% of 39

_______________　　_______________　　_______________　　_______________

5. 43% of 57　　　**6.** 13% of 127　　　**7.** 61% of 535　　　**8.** 53% of 175

_______________　　_______________　　_______________　　_______________

Use 1% or 10% to estimate the percent of each number.

9. 37% of 63　　　**10.** 93% of 57　　　**11.** 16% of 83　　　**12.** 22% of 45

_______________　　_______________　　_______________　　_______________

13. 5% of 91　　　**14.** 13% of 65　　　**15.** 58% of 135　　　**16.** 41% of 543

_______________　　_______________　　_______________　　_______________

Every Tuesday, the Build-It-Yourself Warehouse gives senior citizens a 5% discount. Use estimation to complete the chart below to find the approximate cost of these items purchased on a Tuesday by a senior citizen.

	Item	Original Price	Estimated Discount	Estimated Final Price
17.	Electric drill	$45.99		
18.	Screwdriver set	$17.99		
19.	Rake	$23.50		
20.	Flower box	$37.79		
21.	Power saw	$95.29		
22.	10 lb bag of grass seed	$12.75		

Holt Mathematics

LESSON 6-3 | Reteach
Estimate with Percents

To estimate the percent of a number, choose a fraction that is close
to the given percent.

Percent	5%	10%	20%	25%	$33\frac{1}{3}$	40%	50%	60%	75%	80%
Fraction	$\frac{1}{20}$	$\frac{1}{10}$	$\frac{1}{5}$	$\frac{1}{4}$	$\frac{1}{3}$	$\frac{2}{5}$	$\frac{1}{2}$	$\frac{3}{5}$	$\frac{3}{4}$	$\frac{4}{5}$

Estimate 27% of 123.

27% is about 25%, which is equivalent to $\frac{1}{4}$.

123 rounded to the nearest ten is 120.

$\frac{1}{4} \cdot 120 = 30$

So, 27% of 123 is about 30.

Use a fraction to estimate the percent of each number.

1. 17% of 49

17% is about _____, which is _____.

49 rounded to the nearest ten is _____.

_____ • _____ = _____

So, 17% of 49 is about _____.

2. 58% of 298

58% is about _____, which is _____.

298 rounded to the nearest ten is _____.

_____ • _____ = _____

So, 58% of 298 is about _____.

3. 4% of 42

4% is about _____, which is _____.

42 rounded to the nearest ten is _____.

_____ • _____ = _____

So, 4% of 42 is about _____.

4. 46% of 533

46% is about _____, which is _____.

533 rounded to the nearest ten is _____.

_____ • _____ = _____

So, 46% of 533 is about _____.

5. 11% of 88

____

6. 74% of 203

____

7. 21% of 346

____

8. 79% of 55

____

Holt Mathematics

Reteach
Estimate with Percents (continued)

You can also choose a percent that is close to the given percent and use combinations of smaller percents.

1% of a number	Multiply by 0.01.
5% of a number	Find 10% of the number and divide by 2.
10% of a number	Multiply by 0.1.

Estimate 68% of 383.

68% is about 70%.

70% = 7 • 10%

383 rounded to the nearest ten is 380.

10% of 380 is 38, which is about 40.

7 • 40 = 280

So, 70% of 383 is about 280.

Use 1% or 10% to estimate the percent of each number.

9. 15% of 278

15% = _______ + _______

278 rounded to the nearest ten _______

_______% of 280 = _______.

_______% of 280 = _______.

_______ + _______ = _______

So, 15% of 278 is about _______.

10. 61% of 124

61% is about _______%.

_______% = _______ • _______%

124 rounded to the nearest ten _______.

_______% of _______ = _______.

_______ • _______ = _______

So, 61% of 124 is about _______.

11. 18% of 62   **12.** 42% of 179   **13.** 5% of 317   **14.** 79% of 145

_______   _______   _______   _______

Holt Mathematics

Challenge
The Octagon Garden

Estimate each percent. Find your way through the octagon garden by moving to the neighboring octagon with the next greater value. Keep going until you find your way out.

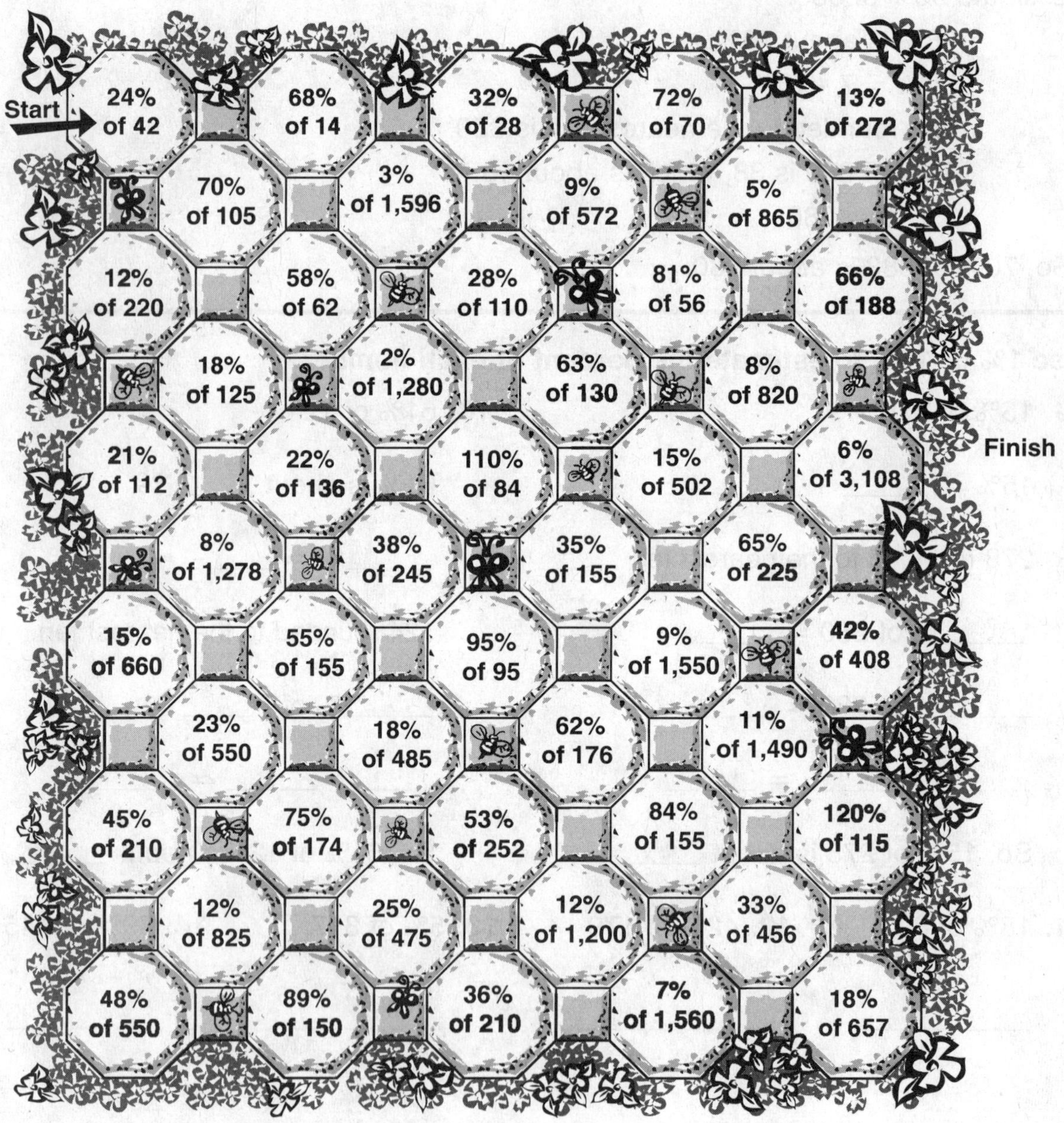

Holt Mathematics

<table>
<tr><td>LESSON
6-3</td><td>

Problem Solving
Estimate with Percents
</td></tr>
</table>

Write the correct answer.

Use the graph to solve Exercises 1–3.

1. If a ski resort in the Rocky Mountain region had 125 visitors, about how many would be snowboarders?

____

2. Recently at one ski resort, 120 out of 400 guests were snowboarders. In which region is this resort?

____

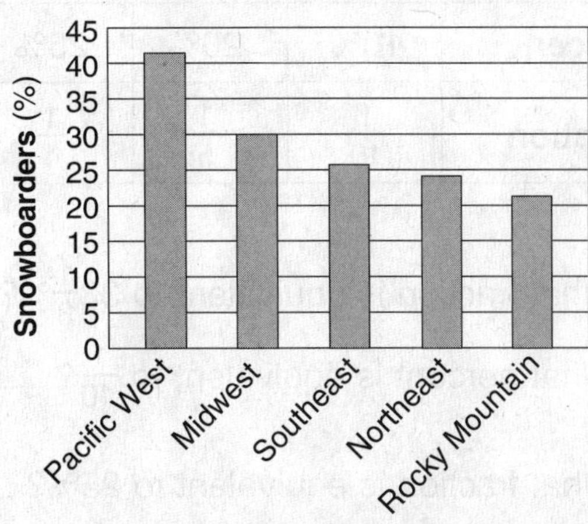

3. Last year, a ski resort in the Northeast region had an average of 556 visitors per weekend. About how many of them were snowboarders?

4. A large department store has an average of 3,456 shoppers per day. About 26% of the shoppers buys something in the store. About how many shoppers each day spend money in the store?

Choose the letter for the best answer.

5. LaToya bought a new car for $29,000. She is entitled to an 11% rebate. About how much will the car cost after the rebate?

  A $3,000   **C** $22,000

  B $10,000   **D** $26,000

6. The cost of a video game is $29.95. Sales tax is 6%. About how much will the video game cost, including tax?

  F about $29   **H** about $32

  G about $30   **J** about $35

7. In 2004, the Brooklyn Public Library spent about $7.6 million on acquisitions. The Detroit Public Library spent about 26% of that amount. About how much money did the Detroit Public Library spend?

  A $2 million   **C** $8 million

  B $3 million   **D** $26 million

8. In 2002, the average starting salary of a high school teacher in Switzerland was $48,704. The average starting salary of a high school teacher in the United States was about 61% of that amount. About how much was the average starting salary in the U.S?

  F $6,000   **H** $30,000

  G $24,000   **J** $60,000

Holt Mathematics

Reading Strategies
Using a Table

The table shows commonly used percents and their fraction equivalents.

Percent	10%	20%	25%	$33\frac{1}{3}$%	50%
Fraction	$\frac{1}{10}$	$\frac{1}{5}$	$\frac{1}{4}$	$\frac{1}{3}$	$\frac{1}{2}$

1. What fraction is equivalent to $33\frac{1}{3}$%? _______________

2. What percent is equivalent to $\frac{1}{10}$? _______________

3. What fraction is equivalent to 25%? _______________

Estimation is used when an exact answer is not needed. Estimating with percents is a useful skill.

Write *exact answer* or *estimate* to show which method is best for the situation.

4. You are trying to decide if you have enough money to buy a new CD.

____

5. You are asked how much you would charge to mow a neighbor's yard.

____

6. You are figuring the discount on a sale item.

____

The table of percent and fraction equivalents can help you make good estimates. Answer each question.

7. What percent is close to 30%? _______________

8. What fraction is close to 45%? _______________

9. What percent is close to 8%? _______________

Holt Mathematics

LESSON 6-3 Puzzles, Twisters & Teasers
Learn All About It!

Choose the better estimate. Then answer the riddle.

S   52% of 234 about   117 or 85.

T   31% of 180 about   36 or 54.

A   18% of 150 about   30 or 50.

E   8% of 310   about   31 or 16.

C   11% of 99   about   15 or 10.

H   23% of 98   about   20 or 30.

Y   8% of 261   about   26 or 16.

W   41% of 16   about   6 or 10.

P   5% of 82   about   8 or 4.

L   52% of 83   about   31 or 41.

O   10% of 48   about   5 or 10.

R   39% of 19   about   4 or 8.

Why did all the cats jump off the fence?

____ ____ ____ ____   ____ ____ ____ ____   ____ ____ ____
54   20   31   26   6   31   8   31   30   41   41

____ ____ ____ ____ ____ ____ ____ ____
10   5   4   26   10   30   54   117

Holt Mathematics

LESSON 6-4 — Practice A
Percent of a Number

45	72	12	27	19	1.2	129	1.9	180
343	128	24	0.19	216	60	81	6	75

Find the percent of each number. Match each answer with a number in the box.

1. 60% of 40

2. 15% of 80

3. 5% of 120

4. 1% of 190

____

____

____

____

5. 120% of 50

6. 43% of 300

7. 150% of 30

8. 36% of 600

____

____

____

____

9. 9% of 800

10. 200% of 64

11. 27% of 300

12. 225% of 12

____

____

____

____

Find the percent of each number. Check whether your answer is reasonable.

13. 20% of 75

14. 25% of 64

15. 4% of 75

16. 125% of 60

____

____

____

____

17. 2% of 400

18. 160% of 80

19. 12% of 50

20. 230% of 70

____

____

____

____

21. 87% of 500

22. 28% of 250

23. 500% of 25

24. 3% of 300

____

____

____

____

25. The highest recorded temperature in Alaska occurred in 1915 at 100°F. The highest recorded temperature in Indiana occurred in 1936 and was 16% higher than the highest temperature in Alaska. What is the highest recorded temperature in Indiana?

____

Holt Mathematics

Practice B
Percent of a Number

Find the percent of each number.

1. 25% of 56   **2.** 10% of 110   **3.** 5% of 150   **4.** 90% of 180

_______   _______   _______   _______

5. 125% of 48   **6.** 225% of 88   **7.** 2% of 350   **8.** 285% of 200

_______   _______   _______   _______

9. 150% of 125   **10.** 46% of 235   **11.** 78% of 410   **12.** 0.5% of 64

_______   _______   _______   _______

Find the percent of each number. Check whether your answer is reasonable.

13. 55% of 900   **14.** 140% of 50   **15.** 75% of 128   **16.** 3% of 600

_______   _______   _______   _______

17. 16% of 85   **18.** 22% of 105   **19.** 0.7% of 110   **20.** 95% of 500

_______   _______   _______   _______

21. 3% of 750   **22.** 162% of 250   **23.** 18% of 90   **24.** 23.2% of 125

_______   _______   _______   _______

25. 0.1% of 950   **26.** 11% of 300   **27.** 52% of 410   **28.** 250% of 12

_______   _______   _______   _______

29. The largest frog in the world is the goliath, found in West Africa.
This type of frog can grow to be 12 inches long. The smallest
frog in the world is about 4% as long as the goliath. What is the
approximate length of the smallest frog in the world?

____

Holt Mathematics

Practice C
Percent of a Number

Find the percent of each number. Round answers to the nearest tenth, if necessary.

1. 67% of 131

2. 19% of 73

3. 7% of 129

4. 1.1% of 67

____

5. 128% of 75

6. 227% of 183

7. 159% of 271

8. 271% of 159

____

9. 17% of 91

10. 23% of 77

11. 7% of 153

12. 7.3% of 73

____

13. 1.95% of 65

14. 137% of 81

15. 0.03% of 532

16. 541% of 132

____

Solve.

17. What number is 55% of 250?

18. 39% of 115 is what number?

____

19. 6% of 95 is what number?

20. What number is 80% of 860?

____

21. What number is 41% of 308?

22. 125% of 425 is what number?

____

23. 300% of 57 is what number?

24. What number is 180% of 110?

____

25. In 2004, there were 10,649 commercial radio stations in the United States. About 19.22% of those stations were devoted to country music. About how many country music radio stations were there in 2004?

____

Holt Mathematics

LESSON 6-4 Reteach
Percent of a Number

You can use this proportion to solve percent problems.

$$\frac{\text{part}}{\text{total}} = \frac{\text{percent}}{100}$$

Find 20% of 65.

$$\frac{x}{65} = \frac{20}{100}$$

$100 \cdot x = 65 \cdot 20$ Find the cross products.

$100x = 1{,}300$

$\frac{100x}{100} = \frac{1{,}300}{100}$ Simplify.

$x = 13$

So, 20% of 65 is 13.

20 is the **percent,** and 65 is the **total.** So, the **part** is unknown.

Find the percent of each number.

1. 40% of 90

   a. part = ____________

   b. total = ____________

   c. percent = ____________

   d. $\dfrac{x}{} = \dfrac{}{100}$

   e. $100x = $ ____________

   f. $x = $ ____________

2. 85% of 520

   a. part = ____________

   b. total = ____________

   c. percent = ____________

   d. $\dfrac{x}{} = \dfrac{}{}$

   e. $100x = $ ____________

   f. $x = $ ____________

3. 30% of 80

4. 47% of 300

5. 45% of 200

6. 120% of 70

________   ________   ________   ________

7. 65% of 40

8. 25% of 76

9. 115% of 40

10. 275% of 12

________   ________   ________   ________

Holt Mathematics

LESSON 6-4 Challenge
Percent Puzzler

Use the percent clues to find each number below.

1. This number is 8 more than $83\frac{1}{2}\%$ of 20. _____________

2. This number is 5 less than 28% of 292. _____________

3. This number is equal to the sum of 4% of 75 and $32\frac{1}{2}\%$ of 64. _____________

4. This number is 7.82 more than $15\frac{3}{4}\%$ of 320. _____________

5. This number is equal to the sum of 52% of 86 and 35% of 52. _____________

6. This number is equal to the product of 8% of 75 and 12% of 60. _____________

7. This number is equal to the sum of 5% of 125 and $55\frac{1}{2}\%$ of 20. _____________

8. This number is equal to the product of 24% of 225 and $3\frac{1}{5}\%$ of 45. _____________

9. This number is equal to the sum of $9\frac{3}{10}\%$ of 110 and $38\frac{1}{4}\%$ of 124. _____________

10. This number is 23.04 less than $14\frac{1}{2}\%$ of 768. _____________

11. This number is 50.21 more than the sum of

    $15\frac{1}{2}\%$ of 42 and 64% of 112. _____________

12. This number is equal to the difference of

    150% of 85 and 280% of 0.3. _____________

13. This number is equal to the square of $7\frac{1}{2}\%$ of 160. _____________

14. This number is 6.8 less than the product of 5 and 8% of 517. _____________

Holt Mathematics

LESSON 6-4 · Problem Solving
Percent of a Number

Write the correct answer.

The world population is estimated to exceed 9 billion by the year 2050. Use the circle graph to solve Exercises 1–3.

1. What is the estimated population of Africa in the year 2050?

   ____

2. Which continent is estimated to have more than 5.31 billion people by the year 2050?

   ____

3. In the year 2002, the world population was estimated at 6 billion people. Based on research from the World Bank, about 20% lived on less than $1 per day. How many people lived on less than $1 per day?

   ____

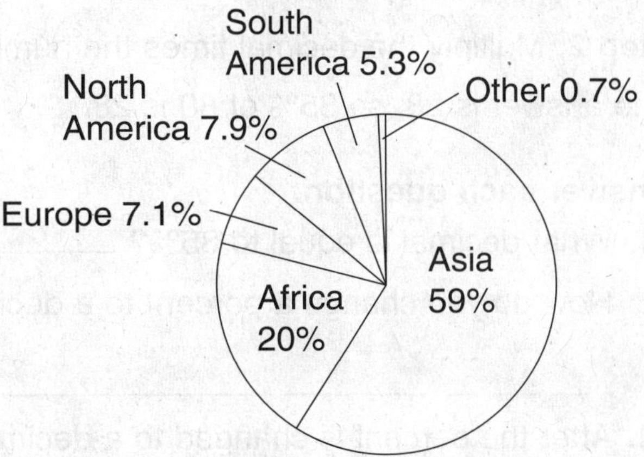

4. What is the combined estimated population for North and South America by the year 2050?

   ____

Choose the letter of the best answer.

5. The student population at King Middle School is 52% female. The total student population is 1,225 students. How many boys go to King Middle School?

   A 538 boys   **C** 637 boys
   B 588 boys   **D** 1,173 boys

6. There are 245 students in the seventh grade. If 40% of them ride the bus to school, how many seventh graders do not ride the bus to school?

   F 98 students   **H** 185 students
   G 147 students   **J** 205 students

7. A half-cup of pancake mix has 5% of the total daily allowance of cholesterol. The total daily allowance of cholesterol is 300 mg. How much cholesterol does a half-cup of pancake mix have?

   A 100 mg   **C** 20 mg
   B 60 mg   **D** 15 mg

8. Carey needs $45 to buy her mother a birthday present. She has saved 22% of the amount so far. How much more does she need?

   F $35.10   **H** $23.00
   G $44.01   **J** $9.90

Holt Mathematics

Reading Strategies
Find a Pattern

You can use decimals to find the percent of a number.

Find 35% of 80.

35% of 80 means 35% times 80.

Step 1: Change the percent to a decimal by moving the decimal point two places to the left. 35% → 0.35

Step 2: Multiply the decimal times the number. → 0.35 × 80

The answer is 28, so 35% of 80 is 28.

Answer each question.

1. What decimal is equal to 35%? _______________

2. How do you change a percent to a decimal?

   ____

3. After the percent is changed to a decimal, what is the next step in finding the percent of a number?

   ____

4. Write 10% as a decimal. _______________

5. What is 10% of 60? _______________

6. What is 20% of 60? _______________

7. What is 30% of 60? _______________

8. What is 40% of 60? _______________

9. What pattern did you notice in the answers to 10, 20, 30, and 40 percent of 60?

   ____

Holt Mathematics

<table>
<tr><td>LESSON
6-4</td><td>

Puzzles, Twisters & Teasers
Hi and Dry!
</td></tr>
</table>

Decide whether or not each equation is correct. Circle the letters above your answers. Then solve the riddle.

1. 7% of 200 = 21

B	**Y**
correct	incorrect

2. 16% of 50 = 18

F	**O**
correct	incorrect

3. 8% of 50 = 4

U	**G**
correct	incorrect

4. 5% of 12 = 0.06

K	**R**
correct	incorrect

5. 67% of 90 = 60.3

S	**L**
correct	incorrect

6. 16% of 70 = 1.17

M	**H**
correct	incorrect

7. 20% of 65 = 15

N	**A**
correct	incorrect

8. 0.5% of 36 = 0.18

D	**P**
correct	incorrect

9. 145% of 210 = 304.5

O	**T**
correct	incorrect

10. 1% of 4 = 4

Z	**W**
correct	incorrect

What can fall on the sea without getting wet?

__ __ __ __ __ __ __

__ __ __ __ __ __

Holt Mathematics

Practice A
Solving Percent Problems

Solve. Choose the letter for the best answer.

1. 50 is 50% of what number?

   A 25   **C** 75

   B 50   **D** 100

2. 25% of what number is 10?

   F 2.5   **H** 40

   G 15   **J** 250

3. What percent of 50 is 10?

   A 5%   **C** 20%

   B 10%   **D** 60%

4. 9 is what percent of 12?

   F 125%   **H** 33%

   G 75%   **J** 30%

5. 25 is 20% of what number?

   A 5   **C** 95

   B 45   **D** 125

6. 75% of what number is 15?

   F 20   **H** 60

   G 25   **J** 90

7. 24 is what percent of 40?

   A 10%   **C** 60%

   B 16%   **D** 64%

8. What percent of 36 is 9?

   F 3%   **H** 45%

   G 25%   **J** 75%

9. What percent of 80 is 48?

____

10. What percent of 50 is 5?

____

11. What percent of 20 is 16?

____

12. What percent of 25 is 15?

____

13. 11 is 50% of what number?

____

14. 40 is 20% of what number?

____

15. 30 is 25% of what number?

____

16. 150 is 200% of what number?

____

17. 60% of what number is 84?

____

18. 125 is 25% of what number?

____

19. The sales tax on a $150 DVD player is $6. What is the sales
tax rate?

____

Holt Mathematics

Practice B
Solving Percent Problems

1. 50 is 40% of what number?

2. 12 is 25% of what number?

3. 18 is what percent of 60?

4. 12 is what percent of 96?

5. 4% of what number is 25?

6. 80% of what number is 160?

7. What percent of 55 is 22?

8. What percent of 75 is 6?

9. 15 is 30% of what number?

10. 8% of what number is 2?

11. 7 is what percent of 105?

12. 24 is 40% of what number?

13. 10% of what number is 14?

14. 16 is what percent of 200?

15. What percent of 32 is 4?

16. What percent of 150 is 60?

17. 1% of what number is 11?

18. 20% of what number is 14?

19. The sales tax on a $750 computer at J & M Computers is $48.75. What is the sales tax rate?

20. A hardcover book sells for $24 at The Bookmart. Ben pays a total of $25.02 for the book. What is the sales tax rate?

Holt Mathematics

LESSON 6-5 Practice C
Solving Percent Problems

Solve. Round answers to the nearest tenth, if necessary.

1. 14 is 28% of what number?

2. 28% of what number is 7?

3. 35% of what number is 70?

4. What percent of 63 is 7?

5. 28 is what percent of 210?

6. 65% of what number is 13?

7. What percent of 180 is 36?

8. 15 is 45% of what number?

9. 7 is 2% of what number?

10. What percent of 144 is 108?

11. What percent of 350 is 840?

12. 13 is what percent of 77?

13. 11 is 4% of what number?

14. 135% of what number is 115?

15. 14 is 16% of what number?

16. 12 is 30% of what number?

17. 18 is what percent of 108?

18. 9 is what percent of 39?

19. I am thinking of a number. This number is 18% of 425. What is the number? _______

20. I am thinking of a percent. This percent times 135 gives 40.5 as the result. What is the percent? _______

21. Jed buys a sweater that has a sale price of $48. Jed gives the sales clerk $60 and receives $7.80 in change. What is the sales tax rate? _______

Holt Mathematics

LESSON 6-5 Reteach
Solving Percent Problems

You can use this proportion to solve percent problems.

$$\frac{\text{part}}{\text{total}} = \frac{\text{percent}}{100}$$

9 is what percent of 12?

$$\frac{9}{12} = \frac{x}{100}$$

$12 \cdot x = 9 \cdot 100$

$12x = 900$

$$\frac{12x}{12} = \frac{900}{12}$$

$x = 75$

So, 9 is 75% of 12.

24 is 30% of what number?

$$\frac{24}{x} = \frac{30}{100}$$

$30 \cdot x = 24 \cdot 100$

$30x = 2{,}400$

$$\frac{30x}{30} = \frac{2{,}400}{30}$$

$x = 80$

So, 24 is 30% of 80.

Solve.

1. What percent of 25 is 14?

   a. part = ______

   b. total = ______

   c. percent = ______

   d. $\dfrac{}{} = \dfrac{x}{100}$

   e. ______ $x =$ ______

   f. $x =$ ______

   g. 14 is ______ of 25.

2. 80% of what number is 16?

   a. part = ______

   b. total = ______

   c. percent = ______

   d. $\dfrac{}{x} = \dfrac{}{100}$

   e. ______ $x =$ ______

   f. $x =$ ______

   g. 16 is 80% of ______.

3. What percent of 20 is 11?

____

4. 18 is 45% of what number?

____

5. 15 is what percent of 5?

____

6. 75% of what number is 105?

____

Holt Mathematics

LESSON 6-5 Challenge
Percentile Rank

Just as there are three quartiles (the lower quartile, the median, and the upper quartile) that divide a data set into four equal groups, there are 99 *percentiles* that divide a data set into 100 groups.

The definition of a percentile is:

$$\text{percentile of score } x = \frac{\text{number of scores less than or equal to score}}{\text{total number of scores}} \cdot 100$$

The frequency table at the right shows the test scores for 28 students.

Find the percentile corresponding to 80.

$$\text{percentile of 80} = \frac{\text{number of scores less than or equal to 80}}{\text{total number of scores}} \cdot 100$$

$$= \frac{14}{28} \times 100 = 0.5 \times 100 = 50$$

So, 80 is the 50th percentile.

Score	Frequency
100	1
95	2
90	5
85	6
80	7
75	3
70	2
65	2

Use the frequency table to find the percentile corresponding to each score. Round your answer to the nearest whole number.

1. 90

2. 70

3. 100

4. 75

5. 95

6. 85

Use the test scores listed below to find the percentile corresponding to each score. Round your answer to the nearest whole number. (*Hint:* Make a frequency table of the scores.)

84, 77, 77, 77, 92, 77, 84, 84, 95, 84, 68, 92, 84, 100, 77, 77, 84, 92, 77, 92, 92, 95, 77, 68, 84, 100, 92, 84, 95, 92

7. 100

8. 95

9. 92

10. 84

11. 77

12. 68

Holt Mathematics

LESSON 6-5 Problem Solving
Solving Percent Problems

Write the correct answer.

1. At one time during 2001, for every 20 copies of *Harry Potter and the Sorcerer's Stone* that were sold, 13.2 copies of *Harry Potter and the Prisoner of Azkaban* were sold. Express the ratio of copies of *The Prisoner of Azkaban* sold to copies of *The Sorcerer's Stone* sold as a percent.

____

2. A souvenir mug sells for $8.00 at a hotel gift store. Kendra gives the clerk $9.00 and receives $0.56 in change. What is the sales tax rate?

____

3. Craig just finished reading 120 pages of his history assignment. If the assignment is 125 pages, what percent has Craig read so far?

____

4. Hal's Sporting Goods had a 1-day sale. The original price of a mountain bike was $325. On sale, it was $276.25. What is the percent reduction for this sale?

____

Choose the correct letter for the best answer.

5. China's area is about 3.7 million square miles. It is on the continent of Asia, which has an area of about 17.2 million square miles. About what percent of the Asian continent does China cover?

 A about 10%   **C** about 17%

 B about 15%   **D** about 22%

6. After six weeks, the tomato plant that was given extra plant food and water was 26 centimeters tall. The tomato plant that was not given any extra plant food was only 74.5% as tall. How tall was the tomato plant that was not given extra plant food?

 F 1.94 cm   **H** 34.89 cm

 G 19.37 cm   **J** 48.5 cm

7. In a survey, 46 people, which was 20% of those surveyed, chose red as their favorite color. How many people were surveyed?

 A 66 people   **C** 460 people

 B 230 people   **D** 920 people

8. Of the 77 billion food and drink cans, bottles, and jars Americans throw away each year, about 65% of them are cans. How many food and drink cans, to the nearest billion, do Americans throw away each year?

 F 12 billion   **H** 50 billion

 G 17 billion   **J** 65 billion

Holt Mathematics

Reading Strategies

LESSON 6-5

Connecting Words and Symbols

You can connect words and symbols to write equations for percent problems.

| 38 | is | 20% | of | what number? | ← words |
| 38 | = | 20% | • | n | ← symbols |

| 9 | is | what percent | of | 45? | ← words |
| 9 | = | n | • | 45 | ← symbols |

| What percent | of | 80 | is | 10? | ← words |
| n | • | 80 | = | 10 | ← symbols |

Answer each question.

1. What is the symbol for the word "of"?

____

2. What symbol means "is"?

____

3. What symbol in the above examples stands for the unknown number or percent?

____

4. Write the equation for this percent problem using symbols: 6 is 10% of what number?

____

5. Write the words for this equation: $27 = n • 30$.

____

6. Write the symbols for this percent problem: What percent of 40 is 30?

____

Holt Mathematics

<table>
<tr><td>LESSON
6-5</td><td></td></tr>
</table>

Puzzles, Twisters & Teasers
Rock On!

Fill in the blanks to complete each statement. Then solve the riddle.

L   25 is _______% of 100

R   27 is _______% of 30

A   105 is _______% of 75

O   16 is _______% of 48

V   56 is 140% of _______

C   40 is _______% of 60

I   10 is _______% of 80

K   9 is _______% of 45

T   6 is 10% of _______

F   7 is 14% of _______

S   30 is 15% of _______

E   18 is _______% of 6

What kind of concert do pebbles enjoy?

____   ____ ____ ____ ____ ____
140   90 33.3 66.7 20

____ ____ ____ ____ ____ ____ ____ ____
50 300 200 60 12.5 40 140 25

Holt Mathematics

Practice A
Percent of Change

Complete the table.

	Problem	Amount of Change	Original Amount	% of Change
1.	25 is decreased to 17			
2.	24 is increased to 36			
3.	50 is decreased to 40			
4.	40 is increased to 56			

Find each percent of change. Round answers to the nearest tenth, if necessary.

5. 60 is decreased to 15 ______________

6. 15 is increased to 21 ______________

7. 12 is increased to 48 ______________

8. 100 is decreased to 25 ______________

9. 60 is decreased to 40 ______________

10. 80 is increased to 152 ______________

11. 15 is increased to 24 ______________

12. 72 is decreased to 64 ______________

13. The Big Bike Shop has in-line skates on sale for 15% off the regular price. A pair of in-line skates usually costs $89. Find the amount of the discount and the sale price.

____

14. In 1935, there were 15,295 banks in the United States. In 2003, there were 9,182 banks. What is the percent of change? Is it a percent increase or decrease?

____

15. A jewelry store is having a going-out-of-business sale. All merchandise is marked 40% off the regular price. The regular price of a watch is $59.95. Find the amount of the discount and the sale price.

____

16. Last year, there were 381 students at Woodland Middle School. This year, there are 419 students. What is the percent of change? Is it a percent increase or decrease?

____

Holt Mathematics

Practice B

LESSON 6-6

Percent of Change

Find each percent of change. Round answers to the nearest tenth, if necessary.

1. 20 is decreased to 11 __________

2. 24 is increased to 30 __________

3. 56 is decreased to 14 __________

4. 25 is increased to 100 __________

5. 18 is increased to 45 __________

6. 90 is decreased to 75 __________

7. 126 is decreased to 48 __________

8. 65 is increased to 144 __________

9. 42 is increased to 72 __________

10. 84 is decreased to 8 __________

11. 95 is increased to 145 __________

12. 248 is decreased to 200 __________

13. 105 is decreased to 32 __________

14. 75 is increased to 350 __________

15. 93 is decreased to 90 __________

16. 16 is decreased to 2 __________

17. A backpack that normally sells for $39 is on sale for 33% off. Find the amount of the discount and the sale price.

____

18. A sporting goods store is having a closeout on a certain style of running shoes. They are marked 55% off the regular price. The regular price is $79.95. Find the amount of the discount and the sale price.

____

19. A gallery owner purchased a very old painting for $3,000. The painting sells at a 325% increase in price. What is the retail price of the painting?

____

20. In August, the Simons' water bill was $48. In September, it was 15% lower. What was the Simons' water bill in September?

____

Holt Mathematics

Practice C
Percent of Change

Find each percent of change or amount of change. Round answers to the nearest tenth, if necessary.

1. 50 is decreased to 19 __________

2. 25 is increased by 80% __________

3. 40 is decreased to 33 __________

4. 75 is increased to 200 __________

5. $80 is increased by 125% __________

6. $14.40 is decreased by 20% __________

7. 11 is decreased by 10% __________

8. 5.7 is increased to 7.8 __________

9. 42.6 is increased to 71.2 __________

10. 98 is increased by 85% __________

11. 145 is decreased to 105 __________

12. 79 is increased to 133 __________

13. 40 is increased by 12.5% __________

14. 119 is increased to 277 __________

15. 537 is decreased to 289 __________

16. 54 is decreased by 75% __________

17. The school bought 6 new computers for the computer lab. Each computer has a retail price of $979. The school received a discount of 12% on each computer. Find the amount of the discount and the final price for all 6 computers.

____

18. Ali paid $85 for an antique clock several years ago. She recently found that it had increased in value 165%. What is the present value of the clock?

____

19. In 1950, there were 40.3 million cars registered in the United States. In 1990, there were 133.7 million cars registered in the United States. What was the percent increase of registered cars from 1950 to 1990?

____

Holt Mathematics

LESSON 6-6 **Reteach**
Percent of Change

A change in a quantity is often described as a percent increase or percent decrease. To calculate a percent increase or decrease, use this equation.

$$\text{percent of change} = \frac{\text{amount of increase or decrease}}{\text{original amount}} \cdot 100$$

Find the percent of change from 28 to 42.

- First, find the amount of the change. $42 - 28 = 14$
- What is the original amount?   28
- Use the equation. $\frac{14}{28} \cdot 100 = 50\%$

An increase from 28 to 42 represents a 50% increase.

Find each percent of change.

1. 8 is increased to 22
amount of change $22 - 8 =$ ______

original amount ______

______ • 100 = ______%

2. 90 is decreased to 81
amount of change $90 - 81 =$ ______

original amount ______

______ • 100 = ______%

3. 125 is increased to 200
amount of change $200 - 125 =$ ______

original amount ______

______ • 100 = ______%

4. 400 is decreased to 60
amount of change $400 - 60 =$ ______

original amount ______

______ • 100 = ______%

5. 64 is decreased to 48

______

6. 140 is increased to 273

______

7. 30 is decreased to 6

______

8. 15 is increased to 21

______

9. 7 is increased to 21

______

10. 320 is decreased to 304

______

Holt Mathematics

LESSON 6-6 Challenge
Double Percent Change

Solve.

1. Enrollment in the school soccer league was 340 last year. This year, it dropped 15%. What is the enrollment this year? If enrollment increases 15% next year, what will the enrollment be? Round to the nearest whole number.

2. Enrollment in the Wilderness Club increased 8% for each of the last two years. Enrollment two years ago was 175. What was the enrollment for each of the following two years? Round to the nearest whole number.

3. The toy store buys stuffed animals from the manufacturer for $2.40 each. The store then sells them at a 96% increase in price. What is the retail price of each animal? If the customer brings in a coupon for 20% off the retail price, how much will the stuffed animal cost?

4. Enrollment in the Jacksonville Township Basketball League was 425 last year. This year, the number enrolled increased 12%. If enrollment drops 12% next year, how many people will be enrolled in the league? Round to the nearest whole number.

5. The manager of a card store buys cards from the manufacturer for $0.65 each. The store regularly sells the cards at a markup of 130%. For a sale, the manager marks down half of the cards 15%. What is the price of the cards that are on sale?

6. Jamal wants to buy a video game. Two stores in the neighborhood offer the game at $49.95. Jamal has a coupon for 18% off the regular price at Fun Electronics. The other store, Gamers, is having a sale with 8% off everything in the store. Jamal also has a coupon for 10% off the sale price at Gamers. At which store should he buy the game? Explain.

Holt Mathematics

LESSON 6-6

Problem Solving
Percent of Change

Write the correct answer.

1. In 2002, U.S. consumers bought about 8.1 million new cars. In 2003, that number decreased by about 6%. To the nearest hundred thousand, or tenth of a million, how many new cars did U.S. consumers buy in 2003?

2. In Union County, Florida, the 1990 census listed the population at 10,252. The 2000 census listed the population as 13,442. What percent increase is this to the nearest tenth of a percent?

3. World production of motor vehicles increased from about 60 million in 2002 to 62 million in 2003. What was the percent increase to the nearest percent?

4. Arthur's dog, Shep, used to weigh 158 pounds. The vet put him on a diet and he lost 13% of his weight. To the nearest pound, how much does Shep weigh now?

5. The number of volunteers rose from 47 on Monday to 64 on Tuesday. What is the percent increase to the nearest tenth of a percent?

6. Coretta's bowling average decreased from 158 to 133. What is the percent decrease to the nearest tenth of a percent?

Choose the correct letter for the best answer.

7. Shandra scored 75 on her first math test. She scored 20% higher on her next math test. What did she score on the second test?

  A 80   **C** 90

  B 85   **D** 95

8. Jim's Gym had income of $20,350 last month. The total increased by $2,000 this month. What was the percent increase to the nearest percent?

  F 11%   **H** 7%

  G 10%   **J** 6%

9. Last year, the average number of absences in school was 8 students per day. This year, the absentee rate is down to 6 students per day. What is the percent decrease in student absences this year?

  A 75%   **C** 25%

  B 33%   **D** 66%

10. During the 2002-2003 ski season $171 million worth of snowboarding equipment was sold. Sales increased by about 15% during the 2003-2004 season. About how much were sales of snowboarding equipment in the 2003-2004 season?

  F $145 million   **H** $186 million

  G $156 million   **J** $197 million

49

Holt Mathematics

LESSON 6-6 Reading Strategies
Analyze Information

Percent can be used to describe change. It is shown as a ratio.

$$\text{Percent of change} = \frac{\text{amount of change}}{\text{original amount}}$$

The following steps describe how the percent of change is figured on a savings account that starts with $50.

This is the original amount in the account:	$50
This is the current amount in the account:	$30
This is the amount that the account decreased by:	$20

$\text{Percent of change} = \dfrac{\$20}{\$50}$ the amount of change over the original amount

Savings went down, so this ratio is ➝ the **percent of decrease** in savings.

1. How much money was placed into the savings account when it opened?

____

2. Did the number of dollars in the account increase or decrease?

____

3. When you are figuring the percent of change, where is the original number placed in the fraction?

____

Use this information for Exercises 4–6: A clothing salesman sold 25 shirts his first day on the job and 45 shirts the second day.

4. What is the original number of shirts he sold?

____

5. How many more shirts did he sell the second day than the first day?

____

6. Write the fraction that shows the amount of change over the original amount.

____

Holt Mathematics

LESSON 6-6 — Puzzles, Twisters & Teasers
Be Patient!

Circle words from the list in the word search. Then find a word that answers the riddle. Circle it and write it on the line.

percent	change	increase	decrease	round
tenth	substitute	decimal	discount	

```
S P E R C E N T E N T H O
R U D E C I M A L U J O D
O P B D I S C O U N T E E
U A R S U N J K O P L R C
N T A S T U P L N E Q X R
D I L K J I N C R E A S E
Q E S D F G T Y H J K O A
P N U H Y T G U W E T R S
B T X O K W I L T C G Y E
N S M I C H A N G E Q W R
```

When do dentists get angry?

When they run out of ___ ___ ___ ___ ___ ___ ___ ___ .

Holt Mathematics

LESSON 6-7

Practice A

Simple Interest

Find each missing value.

1. $p = \$1{,}000$, $r = 5\%$, $t = 2$ years

$I =$ _______ • _______ • _______

$I =$ _______

2. $p = \$600$, $r = 4\%$, $t = 3$ years

$I =$ _______ • _______ • _______

$I =$ _______

3. $I = \$330$, $r = 3\%$, $t = 1$ year

_______ $= p •$ _______ • _______

$p =$ _______

4. $I = \$270$, $r = 5\%$, $t = 3$ years

_______ $= p •$ _______ • _______

$p =$ _______

5. $I = \$600$, $p = \$2{,}500$, $t = 4$ years

_______ $=$ _______ • $r •$ _______

$r =$ _______

6. $I = \$108$, $p = \$900$, $t = 3$ years

_______ $=$ _______ • $r •$ _______

$r =$ _______

7. $p = \$250$, $r = 6\%$, $t = 5$ years

$I =$ _______

8. $p = \$3{,}000$, $r = 7\%$, $t = 4$ years

$I =$ _______

9. $I = \$750$, $r = 4\%$, $t = 5$ years

$p =$ _______

10. $I = \$696$, $r = 3\%$, $t = 4$ years

$p =$ _______

11. $I = \$425$, $p = \$1{,}700$, $t = 5$ years

$r =$ _______

12. $I = \$1{,}680$, $p = \$12{,}000$, $t = 2$ years

$r =$ _______

13. You deposit \$5,000 in an account that earns 5% simple interest. How long will it be before the total amount is \$6,000? _______

14. After 6 years, an account that earns 4% simple interest has earned \$480 in interest. How much was the initial deposit? _______

15. A deposit of \$7,500 earns \$3,900 over a period of 8 years. What is the simple interest rate? _______

16. You deposit \$4,500 in an account that earns 6% simple interest. How much will be in your account after 5 years? _______

Holt Mathematics

Practice B
Simple Interest

Find each missing value.

1. $p = \$1,500$, $r = 5\%$, $t = 3$ years

$I =$ _____________

2. $p = \$6,000$, $r = 4\%$, $t = 2$ years

$I =$ _____________

3. $I = \$30$, $r = 4\%$, $t = 2$ years

$p =$ _____________

4. $I = \$180$, $r = 5\%$, $t = 3$ years

$p =$ _____________

5. $I = \$20$, $p = \$250$, $t = 2$ years

$r =$ _____________

6. $I = \$144$, $p = \$800$, $t = 3$ years

$r =$ _____________

7. $p = \$525$, $r = 3\%$, $t = 1$ year

$I =$ _____________

8. $p = \$3,200$, $r = 6\%$, $t = 4$ years

$I =$ _____________

9. $I = \$450$, $r = 6\%$, $t = 4$ years

$p =$ _____________

10. $I = \$1,440$, $r = 3\%$, $t = 5$ years

$p =$ _____________

11. $I = \$1,275$, $p = \$5,100$, $t = 5$ years

$r =$ _____________

12. $I = \$3,920$, $p = \$14,000$, $t = 4$ years

$r =$ _____________

13. $p = \$1,300$, $r = 4.5\%$, $t = 6$ months

$I =$ _____________

14. $I = \$47.25$, $r = 3.5\%$, $t = 1.5$ years

$p =$ _____________

15. $I = \$891$, $p = \$2,700$, $t = 5.5$ years

$r =$ _____________

16. $I = \$126$, $p = \$400$, $t = 9$ years

$r =$ _____________

17. You deposit $2,500 in an account that earns 4% simple interest. How long will it be before the total amount is $3,000? _____________

18. You deposit $5,000 in account that earns 6.5% simple interest. How much will be in the account after 3 years? _____________

19. A deposit of $10,000 was made to an account the year you were born. After 12 years, the account is worth $16,600. What simple interest rate did the account earn? _____________

20. How long will it take for $6,500 to double at a simple interest rate of 7%? Round to the nearest tenth of a year. _____________

Holt Mathematics

LESSON 6-7 · Practice C
Simple Interest

Complete the table.

	Principal	Interest Rate	Time	Simple Interest
1.	$2,000	3.5%	3 years	
2.	$1,250		4 years	$350
3.	$500	4.5%		$157.50
4.		5%	54 months	$2,115
5.	$12,000		2 years	$1,080
6.	$1,800	7.5%	6 months	
7.		6%	4 years	$73.80
8.	$8,500	6.5%		$6,630
9.		3.5%	5 years	$700
10.	$3,300	4.75%		$313.50
11.	$6,800		16 years	$2,720
12.	$2,400	5.5%	30 months	

Solve.

13. A deposit of $500 in an account earns 6% simple interest. How long will it be before the total amount is $575? ___________

14. What simple interest rate is needed for $1,000 to grow to $1,071.25 in 9 months? ___________

15. A deposit of $3,000 becomes $3,810 after 6 years. What is the simple interest rate on the account? ___________

16. How long will it take for $1,000 to double at a simple interest rate of 5.5%?

____

17. Consuelo deposited an amount of money in a savings account that earned 6.3% simple interest. After 20 years, she had earned $5,922 in interest. What was her initial deposit? ___________

18. A deposit of $2,500 grew to $3,325 after 6 years. What is the final value of a deposit of $7,500 at the same interest rate for the same period of time? ___________

Holt Mathematics

LESSON 6-7

Reteach
Simple Interest

When you put money into a bank account, you may receive simple interest for loaning the bank your money.

$$\text{Interest} = \text{Principal} \cdot \text{Rate} \cdot \text{Time}$$
$$I = p \cdot r \cdot t$$

You can use the expression $\dfrac{I}{p \cdot r \cdot t}$ to solve interest problems.

- To find interest (I), put your finger over I. Perform the operations for letters you see.

- To find principal (p), put your finger over p. Perform the operations for letters you see.

- To find interest rate (r), put your finger over r. Perform the operations for letters you see.

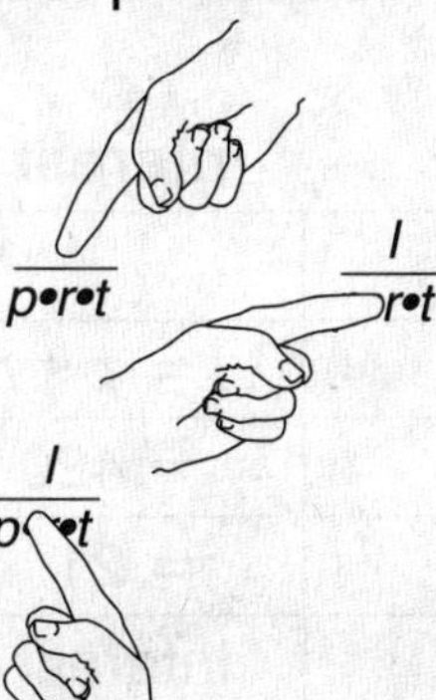

Find each missing value.

1. $p = \$400$, $r = 5\%$, $t = 3$ years

$$I = p \cdot r \cdot t$$

$$I = \underline{\hspace{1cm}} \cdot \underline{\hspace{1cm}} \cdot \underline{\hspace{1cm}}$$

$$I = \underline{\hspace{1cm}}$$

2. $p = \$15,000$, $r = 6\%$, $t = 2$ years

$$I = p \cdot r \cdot t$$

$$I = \underline{\hspace{1cm}} \cdot \underline{\hspace{1cm}} \cdot \underline{\hspace{1cm}}$$

$$I = \underline{\hspace{1cm}}$$

3. $I = \$350$, $r = 7\%$, $t = 2$ years

$$p = \dfrac{I}{r \cdot t}$$

$$p = \dfrac{\underline{\hspace{1cm}}}{\underline{\hspace{1cm}}}$$

$$p = \underline{\hspace{1cm}}$$

4. $I = \$168$, $p = \$1,400$, $t = 4$ years

$$r = \dfrac{I}{p \cdot t}$$

$$r = \dfrac{\underline{\hspace{1cm}}}{\underline{\hspace{1cm}}}$$

$$r = \underline{\hspace{1cm}} = \underline{\hspace{1cm}}$$

5. $I = \$57$, $p = \$380$, $t = 5$ years

$$r = \underline{\hspace{1cm}}$$

6. $p = \$4,800$, $r = 6\%$, $t = 2$ years

$$I = \underline{\hspace{1cm}}$$

7. $I = \$1,200$, $r = 4\%$, $t = 4$ years

$$p = \underline{\hspace{1cm}}$$

8. $p = \$750$, $r = 7\%$, $t = 3$ years

$$I = \underline{\hspace{1cm}}$$

Holt Mathematics

<table>
<tr><td>LESSON
6-7</td><td></td></tr>
</table>

Challenge
Adding On

Simple interest is the amount of interest earned on the original principal. However, you can earn interest on the interest as well as on the principal. This is called **compound interest.**

Josef deposits $400 in a bank that pays 5% interest, compounded annually. Find the amount of interest and principal in Josef's account after 3 years.

Interest and Principal

Year 1	Year 2	Year 3
$I = p \cdot r \cdot t$	$I = p \cdot r \cdot t$	$I = p \cdot r \cdot t$
$= 400 \cdot 0.05 \cdot 1$	$= 420 \cdot 0.05 \cdot 1$	$= 441 \cdot 0.05 \cdot 1$
$= 20$	$= 21$	$= 22.05$
Interest is $20.	Interest is $21.	Interest is $22.05.
Principal is $420.	Principal is $441.	Principal is $463.05.

The amount of interest earned after 3 years is
$20 + $21 + $22.05 = $63.05.

The amount of principal plus interest after 3 years is
$400 + $63.05 = $463.05.

You can also find the total of principal and interest on $400 for 3 years at 5% by multiplying on a calculator
$400 \cdot 1.05 \cdot 1.05 \cdot 1.05$. This shows the amount of interest and principal in Josef's bank account after 3 years, or $463.05.

Find the total amount of interest and principal if interest is compounded annually.

1. $500 for 2 years at 4%

2. $1,200 for 3 years at 4.5%

____

____

3. $300 for 4 years at 5.5%

4. $750 for 2 years at 5.5%

____

____

5. $98 for 3 years at 6.5%

6. $1,056 for 4 years at 5.25%

____

____

7. $520 for 5 years at 4.75%

8. $873 for 6 years at 5.2%

____

____

Holt Mathematics

LESSON 6-7

Problem Solving
Simple Interest

Write the correct answer.

Use the graph to solve Exercises 1–3.

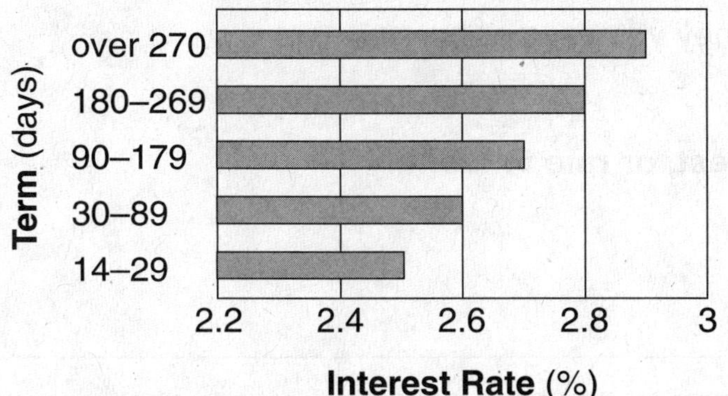

1. How much more interest would be earned on a $100,000 CD for 9 months than for 6 months?

____

2. A customer earned $3,262.50 interest on a 9-month CD. How much was the opening deposit?

____

3. Mrs. Wallace bought a $125,000 CD with a term of 3 years. How much will she earn in 3 years?

____

4. Until June 2002, the simple interest rate on Stafford loans to college students was 5.39% while the student was still in college. How much interest would a student pay on a $1,500 loan for 2 years?

____

5. Diego deposits $4,200 into a savings account that pays 5% simple interest. He decides not to touch the money until it doubles. How long will Diego have to keep the money in this account?

____

Choose the letter of the best answer.

6. Scott took out a 4-year car loan for $5,500. He paid back a total of $7,370. What interest rate did he pay for this loan?

   A 9.5%   **C** 8.5%

   B 9%   **D** 7.5%

7. How much interest would you earn if you were to deposit $575 for 3 months at 2.88% simple interest?

   F $4.14   **H** $41.40

   G $4.83   **J** $48.30

8. How long would you need to keep $775 in an account that pays 3% simple interest to earn $93 interest?

   A 4 years   **C** 4 months

   B 2 years   **D** 2 months

9. If you borrow $12,000 for 30 months at 6.5% simple interest, what is the total amount you will have to repay?

   F $12,065   **H** $13,950

   G $12,780   **J** $21,500

Holt Mathematics

LESSON 6-7
Reading Strategies
Focus on Vocabulary

Principal is the amount of money you save or borrow from a bank.

Interest is the amount of money the bank pays you for the use of your money, or the amount of money you pay the bank to borrow its money.

Rate is the amount of interest paid on money you save or borrow. Rates are usually given as percents.

For Exercises 1–3, write principal, interest, or rate to identify each situation.

1. You have $250 in a savings account.

____

2. The bank pays you 4% a year on the money you have saved.

____

3. The bank paid you $10 on your savings account last year.

____

If you wanted to borrow $3,000 for two years at a rate of 6%, you could use this formula to find the amount of interest you would pay:

$$\text{Interest} = \text{principal} \cdot \text{rate} \cdot \text{time}$$

$$I = p \cdot r \cdot t$$

$$I = \$3,000 \cdot 6\% \cdot 2$$
$$I = \$3,000 \cdot 0.06 \cdot 2 \quad \leftarrow \text{Change percent to decimal.}$$
$$I = \$360 \quad \leftarrow \text{Multiply.}$$

4. What formula is used to find the interest for this loan?

____

5. What is the decimal for 6%?

____

6. How do you find the amount of interest that will be paid on a loan?

____

____

Holt Mathematics

Puzzles, Twisters & Teasers

LESSON 6-7 *Simply Interesting!*

**Use what you know about simple interest to complete the chart.
Then use your answers and the answer key to solve the riddle.**

Principal	Interest Rate	Time	Simple Interest
$3,455	4%	S	$691
F	5.25%	2 years	$630
$16,500	E	24 months	$2,310
A	6%	3 years	$135
$625	3.5%	10 years	C

What do people in clock factories do all day?

They make _______ _______ _______ _______ _______.

$6,000 $750 $218.75   7%   5 years

Holt Mathematics

Practice A
Percents

Write the fraction of the grid that is shaded. Then write the percent.

 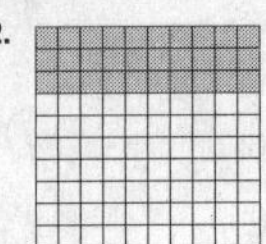 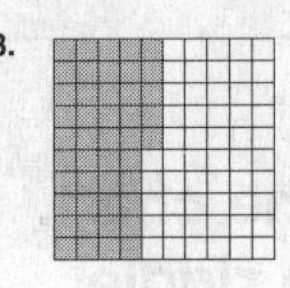

$\frac{30}{100} = 30\%$ $\frac{45}{100} = 45\%$ $\frac{67}{100} = 67\%$

Write each percent as a fraction in simplest form.

4. 20% **5.** 32% **6.** 90% **7.** 43%

$\frac{1}{5}$ $\frac{8}{25}$ $\frac{9}{10}$ $\frac{43}{100}$

8. 48% **9.** 58% **10.** 71% **11.** 80%

$\frac{12}{25}$ $\frac{29}{50}$ $\frac{71}{100}$ $\frac{4}{5}$

12. 28% **13.** 85% **14.** 42% **15.** 52%

$\frac{7}{25}$ $\frac{17}{20}$ $\frac{21}{50}$ $\frac{13}{25}$

Write each percent as a decimal.

16. 19% **17.** 43% **18.** 7% **19.** 62%

0.19   0.43   0.07   0.62

20. 23% **21.** 36% **22.** 59% **23.** 3%

0.23   0.36   0.59   0.03

24. 72.5% **25.** 9.8% **26.** 7.04% **27.** 12.49%

0.725   0.098   0.0704   0.1249

   3   **Holt Mathematics**

Practice B
Percents

Write the percent modeled by each grid.

1. 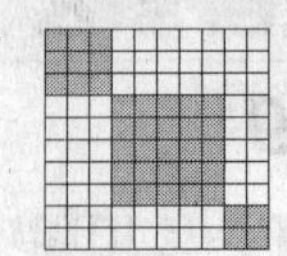 **2.**  **3.** 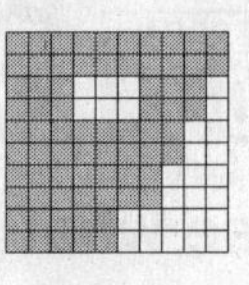

38%   44%   72%

Write each percent as a fraction in simplest form.

4. 16% **5.** 49% **6.** 20% **7.** 15%

$\frac{4}{25}$ $\frac{49}{100}$ $\frac{1}{5}$ $\frac{3}{20}$

8. 18% **9.** 60% **10.** 35% **11.** 46%

$\frac{9}{50}$ $\frac{3}{5}$ $\frac{7}{20}$ $\frac{23}{50}$

12. 86% **13.** 79% **14.** 56% **15.** 45%

$\frac{43}{50}$ $\frac{79}{100}$ $\frac{14}{25}$ $\frac{9}{20}$

Write each percent as a decimal.

16. 33% **17.** 57% **18.** 46% **19.** 6%

0.33   0.57   0.46   0.06

20. 4.7% **21.** 13.2% **22.** 75.8% **23.** 4%

0.047   0.132   0.758   0.04

24. 1.16% **25.** 27.05% **26.** 93.01% **27.** 7.9%

0.0116   0.2705   0.9301   0.079

   4   **Holt Mathematics**

Practice C
Percents

Write the percent modeled by each grid.

1. 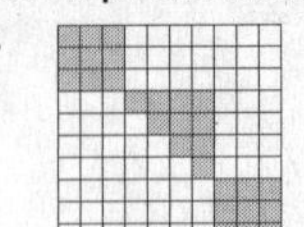 **2.**  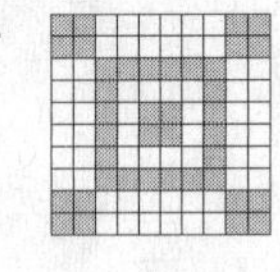 **3.** 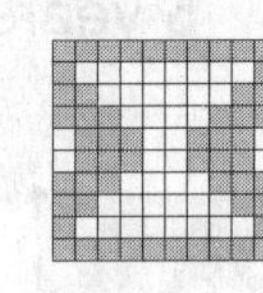

28%   40%   56%

Write each percent as a fraction in simplest form and as a decimal.

4. 29% **5.** 87.5% **6.** 7.5% **7.** 64%

$\frac{29}{100}$; 0.29 $\frac{7}{8}$; 0.875 $\frac{3}{40}$; 0.075 $\frac{16}{25}$; 0.64

8. 62.5% **9.** 34% **10.** 22.5% **11.** 12.5%

$\frac{5}{8}$; 0.625 $\frac{17}{50}$; 0.34 $\frac{9}{40}$; 0.225 $\frac{1}{8}$; 0.125

12. 3.6% **13.** 5.9% **14.** 7.2% **15.** 20.5%

$\frac{9}{250}$; 0.036 $\frac{59}{1000}$; 0.059 $\frac{9}{125}$; 0.072 $\frac{41}{200}$; 0.205

Compare. Write <, >, or =.

16. $\frac{19}{100}$ ▨ 9% **17.** $\frac{44}{100}$ ▨ 44% **18.** $\frac{38}{100}$ ▨ 39%

>   =   <

19. $\frac{12}{25}$ ▨ 40% **20.** $\frac{19}{20}$ ▨ 96% **21.** $\frac{27}{50}$ ▨ 52%

>   <   >

22. $\frac{7}{40}$ ▨ 17.5% **23.** $\frac{5}{12}$ ▨ 41% **24.** $\frac{1}{16}$ ▨ 6.5%

=   >   <

   5   **Holt Mathematics**

Reteach
Percents

To change a percent to a fraction:
• drop the percent symbol;
• write the percent as the numerator of a fraction; $45\% = \frac{45}{100} = \frac{9}{20}$
• write 100 as the denominator;
• simplify.

Write each percent as a fraction in simplest form.

1. $38\% = \frac{38}{100} = \frac{19}{50}$ **2.** $20\% = \frac{20}{100} = \frac{1}{5}$

3. $70\% = \frac{70}{100} = \frac{7}{10}$ **4.** $16\% = \frac{16}{100} = \frac{4}{25}$

5. $36\% = \frac{36}{100} = \frac{9}{25}$ **6.** $8\% = \frac{8}{100} = \frac{2}{25}$

7. $15\% = \frac{3}{20}$ **8.** $53\% = \frac{53}{100}$

9. $24\% = \frac{6}{25}$ **10.** $17\% = \frac{17}{100}$

To change a percent to a decimal:
• drop the percent symbol; $45\% = .45. = 0.45$
• move the decimal point two places to the left. $7\% = .07. = 0.07$

Write each percent as a decimal.

11. 58% **12.** 93% **13.** 15% **14.** 9%

0.58   0.93   0.15   0.09

15. 26% **16.** 2% **17.** 80% **18.** 1%

0.26   0.02   0.8   0.01

19. 23.5% **20.** 9.6% **21.** 40.7% **22.** 7.03%

0.235   0.096   0.407   0.0703

   6   **Holt Mathematics**

      Holt Mathematics

Challenge
Complementary Events

The probability of an event can be expressed as a percent, a fraction, or a decimal.

When two events are the only two events that can occur, they are called **complementary events**. If two events are complementary, the sum of their probabilities is one. For example, suppose the probability of rain is 60%, or 0.6: P(rain) = 0.6. That means the probability of it not raining is 40%, or 0.4: P(not rain) = 0.4. So, P(rain or not rain) = 0.6 + 0.4 = 1.

Draw a line to connect the probability for each Event A to the probability for an Event B that could be a complementary event.

Event A

P (A) = 45%

P (A) = 70%

P (A) = 42%

P (A) = 17.5%

P (A) = 65%

P (A) = 37.5%

P (A) = 48%

Event B

P (B) = 0.29   F

P (B) = $\frac{5}{8}$ T

P (B) = 0.35   L

P (B) = 0.175   N

P (B) = $\frac{13}{25}$ E

P (B) = $\frac{11}{20}$ A

P (B) = $\frac{12}{25}$ I

P (B) = 0.42   S

P (B) = $\frac{3}{10}$ R

P (B) = $\frac{9}{20}$ H

P (B) = $\frac{33}{40}$ U

P (B) = $\frac{29}{50}$ M

Look for the probabilities that do <u>not</u> have lines drawn to them. Use the letters next to those probabilities to write the answer to the riddle. You may use each letter as many times as you wish.

What language do sharks speak?

F I N N I S H

7

Problem Solving
Percents

Write the correct answer.

1. In 2003, 68% of the T.V. set owners in the United States had cable television. Write this percent as a fraction in simplest form and as a decimal.

$\frac{17}{25}$; 0.68

2. In 2004, 27% of Internet users were in the United States. What percent of Internet users were in countries other than the United States?

73%

3. In a survey, 46% of men said they spend fewer than 5 hours shopping for gifts for the holidays. Write this percent as a fraction in simplest form and as a decimal.

$\frac{23}{50}$; 0.46

4. In a survey, 59% of a group of people aged 18-29 said that they do not have enough time to do what they want. What percent of those surveyed feel that they do have enough time to do what they want? Write your answer as a percent and as a decimal.

41%; 0.41

Choose the letter for the best answer.
The table shows the percent of adults who participated in selected leisure activities two or more times per week.

5. Express the percent of adults who dined out two or more times per week as a fraction.

A $\frac{1}{100}$ C $\frac{1}{10}$

B $\frac{1}{50}$ D $\frac{1}{5}$

6. Express the percent of adults who did crossword puzzles two or more times per week as a decimal.

F 0.07   H 0.5

G 0.05   J 0.7

7. What fraction of adults played video games fewer than two times per week?

A $\frac{1}{20}$ C $\frac{19}{50}$

B $\frac{1}{50}$ D $\frac{19}{20}$

Adult Participation in Selected Leisure Activities in 2003	
Activity	**Percent**
Crossword puzzles	7%
Dining out	10%
Reading books	21%
Surfing the net	18%
Video games	5%

8. Which decimal represents the percent of adults who surfed the net fewer than two times per week?

F 0.018   H 0.18

G 0.05   J 0.82

8

Reading Strategies
Multiple Representations

Percent means "per hundred." You can use a hundred grid to picture percents.

The shaded part of the grid is 20% → Read: "20 percent."

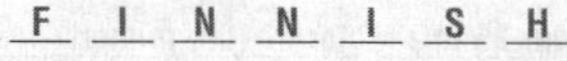

Percents can be written as fractions and as decimals.

20% means "20 per hundred" or $\frac{20}{100}$. The fraction can be written in simplest form. → $\frac{1}{5}$

$\frac{20}{100}$ is read → "20 hundredths" and can be written → 0.20.

The shaded part of the hundred grid can be represented in these ways:

20% → $\frac{20}{100}$ → $\frac{1}{5}$ → 20 hundredths → 0.20

Answer the following questions.

1. What does the word "percent" mean?   per hundred

2. What two other ways can you write percents?   as fractions and as decimals

3. Shade 40% of the hundred grid.

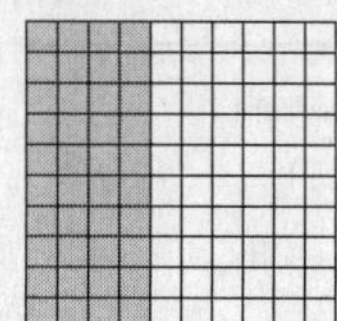

4. Write the shaded part as a decimal.   0.40

5. Write the shaded part as a fraction. $\frac{40}{100}$ or $\frac{2}{5}$

6. If $\frac{1}{2}$ of the grid were shaded, what percent would that be?   50%

9

Puzzles, Twisters, & Teasers
Now You See It...

Write each percent as a fraction in simplest form.

1. 65%   O $\frac{13}{20}$

2. 20%   E $\frac{1}{5}$

3. 72%   T $\frac{18}{25}$

4. 63%   R $\frac{63}{100}$

5. 6%   V $\frac{3}{50}$

6. 24%   K $\frac{6}{25}$

Write each percent as a decimal.

7. 3.62%   L   0.0362

8. 60%   A   0.6

9. 52%   I   0.52

10. 6.3%   M   0.063

11. 45%   P   0.45

12. 36.2%   D   0.362

Write the letter on the line above the correct answer to solve the riddle.

What does the Invisible Man drink at snack times?

He drinks E V A P O R A T E D

$\frac{1}{5}$ $\frac{3}{50}$ 0.6   0.45 $\frac{13}{20}$ $\frac{63}{100}$ 0.6 $\frac{18}{25}$ $\frac{1}{5}$ 0.362

M I L K

0.063   0.52   0.0362 $\frac{6}{25}$

10

Practice A
Fractions, Decimals, and Percents

Complete.

1. $0.16 = \frac{16}{100} = 16\ \%$

2. $0.09 = \frac{9}{100} = 9\ \%$

3. $\frac{3}{5} = \frac{60}{100} = 60\ \%$

4. $\frac{21}{25} = \frac{84}{100} = 84\ \%$

Write each decimal as a percent.

5. 0.34   __34%__
6. 0.08   __8%__
7. 0.19   __19%__
8. 0.64   __64%__

9. 0.561   __56.1%__
10. 0.049   __4.9%__
11. 0.105   __10.5%__
12. 0.9   __90%__

Write each fraction as a percent.

13. $\frac{1}{4}$ __25%__
14. $\frac{7}{50}$ __14%__
15. $\frac{15}{16}$ __93.75%__
16. $\frac{9}{20}$ __45%__

17. $\frac{7}{8}$ __87.5%__
18. $\frac{9}{40}$ __22.5%__
19. $\frac{11}{50}$ __22%__
20. $\frac{4}{5}$ __80%__

Decide whether to use pencil and paper, mental math, or a calculator. Then solve.

21. In a survey, 25 students were asked whether they prefer orange juice or grapefruit juice. Seventeen students said they prefer orange juice. What percent of the students surveyed said they prefer orange juice?

__Possible answer: Since the denominator 25 is a factor of 100, mental math is a good choice; 68%.__

11
Holt Mathematics

Practice B
Fractions, Decimals, and Percents

Write each decimal as a percent.

1. 0.17   __17%__
2. 0.56   __56%__
3. 0.04   __4%__
4. 0.7   __70%__

5. 0.025   __2.5%__
6. 0.803   __80.3%__
7. 0.3   __30%__
8. 0.072   __7.2%__

Write each fraction as a percent.

9. $\frac{13}{40}$ __32.5%__
10. $\frac{3}{5}$ __60%__
11. $\frac{3}{20}$ __15%__
12. $\frac{5}{12}$ __41.7%__

13. $\frac{5}{16}$ __31.25%__
14. $\frac{3}{80}$ __3.75%__
15. $\frac{5}{6}$ __83.3%__
16. $\frac{19}{25}$ __76%__

Decide whether pencil and paper, mental math, or a calculator is most useful when solving the following problems. Then solve.

17. In a survey, 60 baseball fans were asked whether they thought the designated hitter rule should be changed. Forty-one fans thought the rule should be changed. What percent of the fans surveyed said that the designated hitter rule should be changed?

__Possible answer: Since 41 ÷ 60 does not divide evenly, pencil and paper is not a good choice. Since the denominator 60 is not a factor of 100, mental math is not a good choice. So using a calculator is the best method; 68.3%.__

18. The police use a speed gun to monitor one part of a highway. During one hour, 6 out of 25 cars were traveling above the speed limit. What percent of the cars were traveling above the speed limit?

__Possible answer: Since the denominator 25 is a factor of 100, mental math is a good choice; 24%.__

12
Holt Mathematics

Practice C
Fractions, Decimals, and Percents

Write each decimal as a percent.

1. 0.04   __4%__
2. 0.975   __97.5%__
3. 0.031   __3.1%__
4. 0.409   __40.9%__

5. 0.378   __37.8%__
6. 0.4   __40%__
7. 0.524   __52.4%__
8. 0.067   __6.7%__

Write each fraction as a percent.

9. $\frac{13}{80}$ __16.25%__
10. $\frac{7}{9}$ __77.8%__
11. $\frac{17}{20}$ __85%__
12. $\frac{5}{8}$ __62.5%__

13. $\frac{39}{40}$ __97.5%__
14. $\frac{4}{11}$ __36.4%__
15. $\frac{11}{25}$ __44%__
16. $\frac{9}{50}$ __18%__

Compare. Write <, >, or =.

17. 30% __<__ $\frac{1}{3}$
18. $\frac{3}{5}$ __=__ 60%
19. $\frac{7}{16}$ __<__ 44%

20. 0.065 __<__ 65%
21. 0.507 __>__ 50%
22. $\frac{7}{40}$ __<__ 20%

Decide whether pencil and paper, mental math, or a calculator is most useful when solving the following problem. Then solve.

23. In a survey, 80 students were asked to name their favorite subject. Thirty students said that English was their favorite. What percent of the students surveyed said that English was their favorite subject?

Possible answer: Since $\frac{30}{80} = \frac{3}{8}$, and 3 ÷ 8 divides evenly,

__pencil and paper is a good choice; 37.5%.__

13
Holt Mathematics

Reteach
Fractions, Decimals, and Percents

To change a decimal to a percent:
• move the decimal point two places to the right; $0.07 = .07. = 7\%$
• write the % symbol after the number.

Write each decimal as a percent.

1. 0.34   __34%__
2. 0.06   __6%__
3. 0.93   __93%__
4. 0.57   __57%__

5. 0.8   __80%__
6. 0.734   __73.4%__
7. 0.082   __8.2%__
8. 0.225   __22.5%__

9. 0.604   __60.4%__
10. 0.09   __9%__
11. 0.518   __51.8%__
12. 0.039   __3.9%__

To change a fraction to a percent:
• Find an equivalent fraction with a denominator of 100.
• Use the numerator of the equivalent fraction as the percent.

$$\frac{8}{25} = \frac{x}{100}$$
$$\frac{8 \cdot 4}{25 \cdot 4} = \frac{32}{100}$$
$$\frac{8}{25} = \frac{32}{100} = 32\%$$

Think: $100 \div 25 = 4$. So, multiply the numerator and denominator by 4.

Write each fraction as a percent.

13. $\frac{3}{10}$ __30%__
14. $\frac{2}{50}$ __4%__
15. $\frac{7}{20}$ __35%__
16. $\frac{1}{5}$ __20__

17. $\frac{1}{8}$ __12.5%__
18. $\frac{3}{25}$ __12%__
19. $\frac{3}{4}$ __75%__
20. $\frac{23}{40}$ __57.5%__

21. $\frac{11}{20}$ __55%__
22. $\frac{43}{50}$ __86%__
23. $\frac{24}{25}$ __96%__
24. $\frac{7}{8}$ __87.5%__

14
Holt Mathematics

Holt Mathematics

LESSON 6-2
Challenge
Climbing a Percent Pyramid

Begin by changing each ratio in the pyramid to a percent. Then climb to the top by comparing the percents as you go.

The percents must increase as you climb to the top. Each brick on your path must touch the previous brick. Circle the ratios in the path that leads to the top.

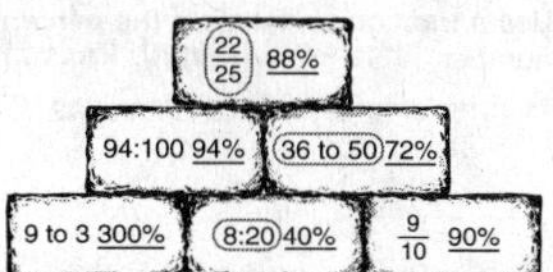

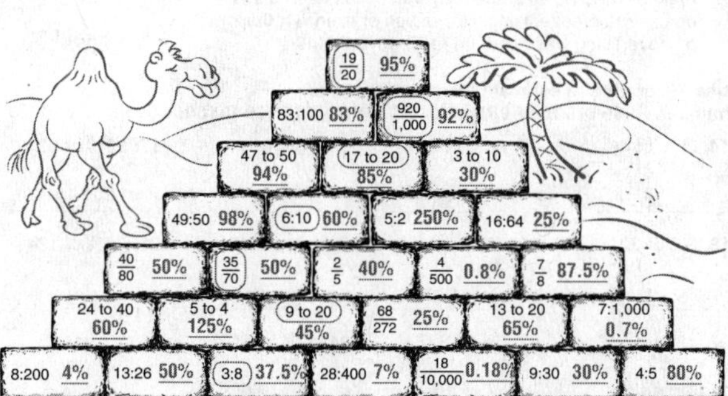

Possible path shown.

15   **Holt Mathematics**

LESSON 6-2
Problem Solving
Fractions, Decimals, and Percents

Write the correct answer.

1. About 543 out of every 1,000 people in the United States owned a cell phone in 2003. In Japan, the rate was 68 for every 100 people. How much greater was the percent of cell phone ownership in Japan than in the U.S.?

   13.7%

2. In 2002, the adult population of the United States was about 206 million. About 113 million people participated in an exercise program. To the nearest percent, what percent of the adult population participated in an exercise program?

   55%

3. When asked about their favorite Thanksgiving leftover, $\frac{1}{20}$ of the people said vegetables and $\frac{7}{100}$ said mashed potatoes. Which food was more popular and by what percent?

   mashed potatoes, 2%

4. In a survey, 80 people were asked whether they thought the speed limit for interstate highways should be raised. Twenty-five people said the speed limit should be raised. What percent of people did not think that the speed limit should be raised?

   68.75%

Choose the letter for the best answer.

The table shows the number of students in four schools who own computers.

5. What percent of students at Percy owns computers?

   A 125%   C 250%
   B 12.5%   **D** 25%

6. In which school does about 73% of the students own computers?

   F Hunter   H Percy
   G Madison   **J** King

7. What percent of students at Madison do not own computers? Round to the nearest tenth of a percent.

   A 3.3%   C 33.3%
   B 9.9%   **D** 66.7% [d]

8. Which school has the greatest percent of students who own computers?

   F Hunter   H Percy
   G Madison   **J** King

Students Who Own Computers

School	Number of Students
Madison	90 out of 270
Hunter	56 out of 100
King	110 out of 150
Percy	125 out of 500

16   **Holt Mathematics**

LESSON 6-2
Reading Strategies
Use a Diagram

You can use a diagram to understand the relationships among fractions, decimals, and percents.

Write 120 out of 400 as a fraction in tenths, as a decimal, and as a percent.

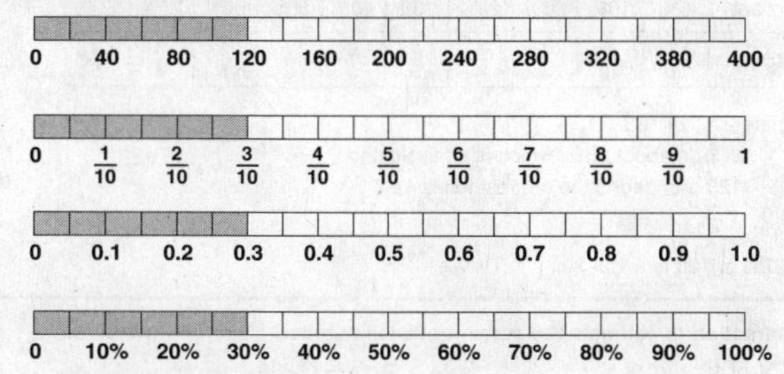

The diagram shows that 120 out of 400 can be written as $\frac{3}{10}$, 0.3, or 30%.

Use the diagram to answer each question.

1. What fraction in tenths can you write for 280 out of 400?

   $\frac{7}{10}$

2. What percent can you write for 280 out 400?

   70%

3. What decimal can you write for 180 out of 400?

   0.45

4. What percent can you write for 180 out 400?

   45%

5. How could you change or add to the diagram above so that you could use it to find the percent that is equal to 110 out of 200?

   Possible answer: Change the labels on the first bar so that the bar shows a total of 200, and each square represents 10 units. Then find the percent that is equivalent to 110 units.

17   **Holt Mathematics**

LESSON 6-2
Puzzles, Twisters, & Teasers
Pick a Percent

Draw a line from each fraction or decimal to the equivalent percent.

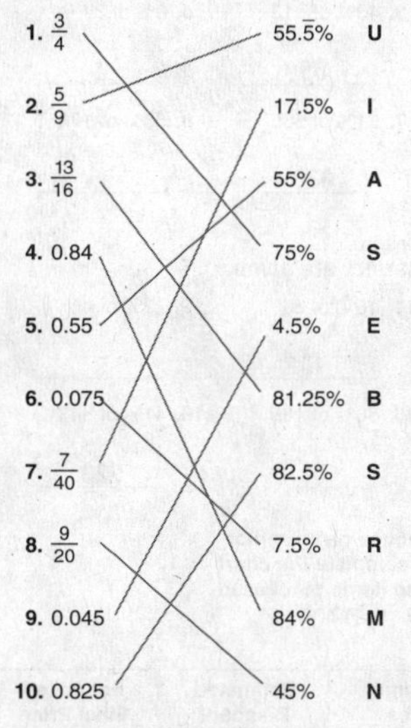

1. $\frac{3}{4}$ 55.$\overline{5}$%   U
2. $\frac{5}{9}$ 17.5%   I
3. $\frac{13}{16}$ 55%   A
4. 0.84   75%   S
5. 0.55   4.5%   E
6. 0.075   81.25%   B
7. $\frac{7}{40}$ 82.5%   S
8. $\frac{9}{20}$ 7.5%   R
9. 0.045   84%   M
10. 0.825   45%   N

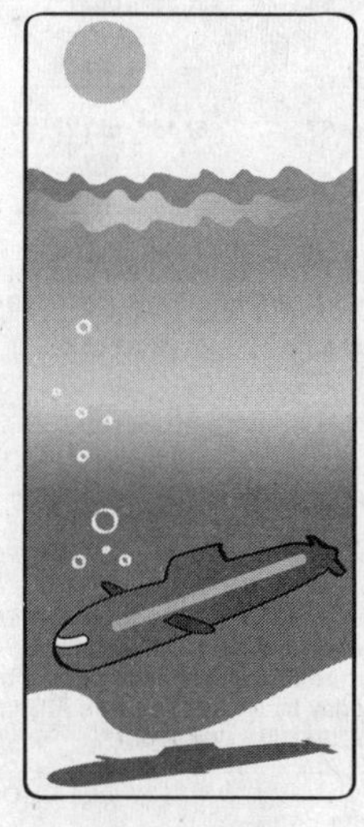

Write the letter on the line above the matching problem number to solve the riddle.

What kind of sandwiches do sailors eat?

S U B M A R I N E S
1. 2. 3. 4. 5. 6. 7. 8. 9. 10.

18   **Holt Mathematics**

Practice A
LESSON 6-3 — *Estimate with Percents*

Which fraction is the best estimate for the given percent? Choose the letter of the best answer.

1. 24% A $\frac{1}{2}$ B $\frac{1}{3}$ Ⓒ $\frac{1}{4}$ D $\frac{1}{5}$

2. 73% F $\frac{4}{5}$ Ⓖ $\frac{3}{4}$ H $\frac{2}{3}$ J $\frac{1}{2}$

3. 51% Ⓐ $\frac{1}{2}$ B $\frac{2}{3}$ C $\frac{3}{4}$ D $\frac{4}{5}$

4. 32% F $\frac{1}{5}$ G $\frac{1}{4}$ Ⓗ $\frac{1}{3}$ J $\frac{1}{2}$

5. 39% A $\frac{1}{5}$ Ⓑ $\frac{2}{5}$ C $\frac{3}{5}$ D $\frac{4}{5}$

6. 12% F $\frac{1}{3}$ G $\frac{1}{4}$ H $\frac{1}{5}$ Ⓙ $\frac{1}{10}$

7. 19% A $\frac{1}{2}$ B $\frac{1}{3}$ C $\frac{1}{4}$ Ⓓ $\frac{1}{5}$

Use a fraction to estimate the percent of each number. Answers may vary. Possible answers given.

8. 24% of 80	9. 73% of 44	10. 51% of 120	11. 32% of 90
20	33	60	30

12. 39% of 50	13. 12% of 40	14. 19% of 30	15. 58% of 200
20	4	6	120

Use 1% or 10% to estimate the percent of each number. Answers may vary. Possible answers given.

16. 39% of 25	17. 11% of 20	18. 3% of 200	19. 71% of 50
10	2	6	35

20. 9% of 40	21. 49% of 60	22. 4% of 50	23. 89% of 30
4	30	2	27

 19 **Holt Mathematics**

Practice B
LESSON 6-3 — *Estimate with Percents*

Use a fraction to estimate the percent of each number. Answers will vary. Possible answers are given.

1. 21% of 82	2. 35% of 42	3. 47% of 164	4. 9% of 68
16	14	82	7

5. 65% of 78	6. 11% of 92	7. 26% of 124	8. 89% of 51
50	9	30	45

9. 77% of 198	10. 5% of 75	11. 31% of 148	12. 53% of 539
150	4	45	270

13. In 2004, about $38 out of every $100 spent on advertising was spent on television advertising. The amount spent on radio advertising was about 21% as much as was spent on television advertising. How much of every $100 spent on advertising was spent on radio advertising? **about $8**

Use 1% or 10% to estimate the percent of each number. Answers may vary. Possible answers are given.

14. 32% of 46	15. 81% of 36	16. 15% of 44	17. 21% of 62
15	32	6	12

18. 3% of 72	19. 62% of 88	20. 12% of 48	21. 65% of 124
2	54	5	84

22. 18% of 147	23. 5% of 837	24. 37% of 213	25. 2% of 188
30	42	84	4

26. The Fresh Acres Swim Club has a $35,000 budget for pool maintenance this year. The club members have agreed to raise the budget by 4%. Estimate the pool maintenance budget for next year. **$36,400**

 20 **Holt Mathematics**

Practice C
LESSON 6-3 — *Estimate with Percents*

Use a fraction to estimate the percent of each number. Answers may vary. Possible answers are given.

1. 23% of 53	2. 35% of 37	3. 46% of 113	4. 8% of 39
13	12	55	4

5. 43% of 57	6. 13% of 127	7. 61% of 535	8. 53% of 175
24	13	330	90

Use 1% or 10% to estimate the percent of each number. Answers may vary. Possible answers are given.

9. 37% of 63	10. 93% of 57	11. 16% of 83	12. 22% of 45
24	54	16	9

13. 5% of 91	14. 13% of 65	15. 58% of 135	16. 41% of 543
5	7	81	220

Every Tuesday, the Build-It-Yourself Warehouse gives senior citizens a 5% discount. Use estimation to complete the chart below to find the approximate cost of these items purchased on a Tuesday by a senior citizen. Answers will vary. Possible answers are given.

	Item	Original Price	Estimated Discount	Estimated Final Price
17.	Electric drill	$45.99	$2.30	$43.70
18.	Screwdriver set	$17.99	$1.00	$17.00
19.	Rake	$23.50	$1.20	$22.30
20.	Flower box	$37.79	$2.00	$36.00
21.	Power saw	$95.29	$5.00	$90.00
22.	10 lb bag of grass seed	$12.75	$0.70	$12.00

 21 **Holt Mathematics**

Reteach
LESSON 6-3 — *Estimate with Percents*

To estimate the percent of a number, choose a fraction that is close to the given percent.

Percent	5%	10%	20%	25%	$33\frac{1}{3}$	40%	50%	60%	75%	80%
Fraction	$\frac{1}{20}$	$\frac{1}{10}$	$\frac{1}{5}$	$\frac{1}{4}$	$\frac{1}{3}$	$\frac{2}{5}$	$\frac{1}{2}$	$\frac{3}{5}$	$\frac{3}{4}$	$\frac{4}{5}$

Estimate 27% of 123.

27% is about 25%, which is equivalent to $\frac{1}{4}$.

123 rounded to the nearest ten is 120.

$\frac{1}{4} \cdot 120 = 30$

So, 27% of 123 is about 30.

Use a fraction to estimate the percent of each number. Possible answers:

1. 17% of 49

17% is about **20%**, which is $\frac{1}{5}$.

49 rounded to the nearest ten is **50**.

$\frac{1}{5} \cdot 50 = 10$

So, 17% of 49 is about **10**.

2. 58% of 298

58% is about **60%**, which is $\frac{3}{5}$.

298 rounded to the nearest ten is **300**.

$\frac{3}{5} \cdot 300 = 180$

So, 58% of 298 is about **180**.

3. 4% of 42

4% is about **5%**, which is $\frac{1}{20}$.

42 rounded to the nearest ten is **40**.

$\frac{1}{20} \cdot 40 = 2$

So, 4% of 42 is about **2**.

4. 46% of 533

46% is about **50%**, which is $\frac{1}{2}$.

533 rounded to the nearest ten is **530**.

$\frac{1}{2} \cdot 530 = 265$

So, 46% of 533 is about **265**.

5. 11% of 88	6. 74% of 203	7. 21% of 346	8. 79% of 55
9	150	70	48

 22 **Holt Mathematics**

 64 **Holt Mathematics**

LESSON **6-3**

Reteach
Estimate with Percents (continued)

You can also choose a percent that is close to the given percent and use combinations of smaller percents.

1% of a number	Multiply by 0.01.
5% of a number	Find 10% of the number and divide by 2.
10% of a number	Multiply by 0.1.

Estimate 68% of 383.

68% is about 70%.

70% = 7 • 10%

383 rounded to the nearest ten is 380.

10% of 380 is 38, which is about 40.

7 • 40 = 280

So, 70% of 383 is about 280.

Use 1% or 10% to estimate the percent of each number. Possible answers:

9. 15% of 278

15% = 10% + 5%

278 rounded to the nearest ten 280

10 % of 280 = 28.

5 % of 280 = 14.

28 + 14 = 42

So, 15% of 278 is about 42.

10. 61% of 124

61% is about 60 %.

60 % = 6 • 10 %

124 rounded to the nearest ten 120.

10 % of 120 = 12.

6 • 12 = 72

So, 61% of 124 is about 72.

11. 18% of 62 — 12

12. 42% of 179 — 72

13. 5% of 317 — 16

14. 79% of 145 — 120

LESSON **6-3**

Challenge
The Octagon Garden

Estimate each percent. Find your way through the octagon garden by moving to the neighboring octagon with the next greater value. Keep going until you find your way out.

Estimates may vary.

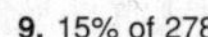

LESSON **6-3**

Problem Solving
Estimate with Percents

Write the correct answer. Answers will vary. Possible answers are given.

Use the graph to solve Exercises 1–3.

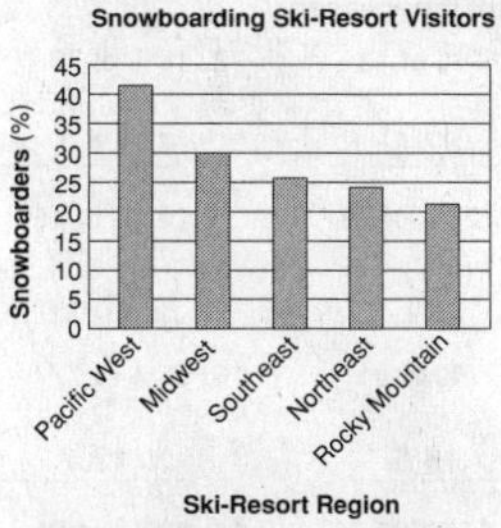

1. If a ski resort in the Rocky Mountain region had 125 visitors, about how many would be snowboarders?

about 25 visitors

2. Recently at one ski resort, 120 out of 400 guests were snowboarders. In which region is this resort?

Midwest region

3. Last year, a ski resort in the Northeast region had an average of 556 visitors per weekend. About how many of them were snowboarders?

about 140 visitors

4. A large department store has an average of 3,456 shoppers per day. About 26% of the shoppers buys something in the store. About how many shoppers each day spend money in the store?

about 850 shoppers

Choose the letter for the best answer.

5. LaToya bought a new car for $29,000. She is entitled to an 11% rebate. About how much will the car cost after the rebate?

A $3,000 **C** $22,000
B $10,000 **(D)** $26,000

6. The cost of a video game is $29.95. Sales tax is 6%. About how much will the video game cost, including tax?

F about $29 **(H)** about $32
G about $30 **J** about $35

7. In 2004, the Brooklyn Public Library spent about $7.6 million on acquisitions. The Detroit Public Library spent about 26% of that amount. About how much money did the Detroit Public Library spend?

(A) $2 million **C** $8 million
B $3 million **D** $26 million

8. In 2002, the average starting salary of a high school teacher in Switzerland was $48,704. The average starting salary of a high school teacher in the United States was about 61% of that amount. About how much was the average starting salary in the U.S?

F $6,000 **(H)** $30,000
G $24,000 **J** $60,000

LESSON **6-3**

Reading Strategies
Using a Table

The table shows commonly used percents and their fraction equivalents.

Percent	10%	20%	25%	$33\frac{1}{3}$%	50%
Fraction	$\frac{1}{10}$	$\frac{1}{5}$	$\frac{1}{4}$	$\frac{1}{3}$	$\frac{1}{2}$

1. What fraction is equivalent to $33\frac{1}{3}$%? $\frac{1}{3}$

2. What percent is equivalent to $\frac{1}{10}$? 10%

3. What fraction is equivalent to 25%? $\frac{1}{4}$

Estimation is used when an exact answer is not needed. Estimating with percents is a useful skill.

Write *exact answer* or *estimate* to show which method is best for the situation.

4. You are trying to decide if you have enough money to buy a new CD.

estimate

5. You are asked how much you would charge to mow a neighbor's yard.

exact answer

6. You are figuring the discount on a sale item.

estimate

The table of percent and fraction equivalents can help you make good estimates. Answer each question.

7. What percent is close to 30%? $33\frac{1}{3}$%

8. What fraction is close to 45%? $\frac{1}{2}$

9. What percent is close to 8%? 10%

Puzzles, Twisters & Teasers
Learn All About It!

Choose the better estimate. Then answer the riddle.

S 52% of 234 about (117) or 85.

T 31% of 180 about 36 or (54).

A 18% of 150 about (30) or 50.

E 8% of 310 about (31) or 16.

C 11% of 99 about 15 or (10).

H 23% of 98 about (20) or 30.

Y 8% of 261 about (26) or 16.

W 41% of 16 about (6) or 10.

P 5% of 82 about 8 or (4).

L 52% of 83 about 31 or (41).

O 10% of 48 about (5) or 10.

R 39% of 19 about 4 or (8).

Why did all the cats jump off the fence?

T	H	E	Y		W	E	R	E		A	L	L
54	20	31	26		6	31	8	31		30	41	41

C	O	P	Y	C	A	T	S
10	5	4	26	10	30	54	117

  Holt Mathematics

Practice A
Percent of a Number

45	72	12	27	19	1.2	129	1.9	180
343	128	24	0.19	216	60	81	6	75

Find the percent of each number. Match each answer with a number in the box.

1. 60% of 40

    24

2. 15% of 80

    12

3. 5% of 120

    6

4. 1% of 190

    1.9

5. 120% of 50

    60

6. 43% of 300

    129

7. 150% of 30

    45

8. 36% of 600

    216

9. 9% of 800

    72

10. 200% of 64

    128

11. 27% of 300

    81

12. 225% of 12

    27

Find the percent of each number. Check whether your answer is reasonable.

13. 20% of 75

    15

14. 25% of 64

    16

15. 4% of 75

    3

16. 125% of 60

    75

17. 2% of 400

    8

18. 160% of 80

    128

19. 12% of 50

    6

20. 230% of 70

    161

21. 87% of 500

    435

22. 28% of 250

    70

23. 500% of 25

    125

24. 3% of 300

    9

25. The highest recorded temperature in Alaska occurred in 1915 at 100°F. The highest recorded temperature in Indiana occurred in 1936 and was 16% higher than the highest temperature in Alaska. What is the highest recorded temperature in Indiana?

    116°F

  Holt Mathematics

Practice B
Percent of a Number

Find the percent of each number.

1. 25% of 56

    14

2. 10% of 110

    11

3. 5% of 150

    7.5

4. 90% of 180

    162

5. 125% of 48

    60

6. 225% of 88

    198

7. 2% of 350

    7

8. 285% of 200

    570

9. 150% of 125

    187.5

10. 46% of 235

    108.1

11. 78% of 410

    319.8

12. 0.5% of 64

    0.32

Find the percent of each number. Check whether your answer is reasonable.

13. 55% of 900

    495

14. 140% of 50

    70

15. 75% of 128

    96

16. 3% of 600

    18

17. 16% of 85

    13.6

18. 22% of 105

    23.1

19. 0.7% of 110

    0.77

20. 95% of 500

    475

21. 3% of 750

    22.5

22. 162% of 250

    405

23. 18% of 90

    16.2

24. 23.2% of 125

    29

25. 0.1% of 950

    0.95

26. 11% of 300

    33

27. 52% of 410

    213.2

28. 250% of 12

    30

29. The largest frog in the world is the goliath, found in West Africa. This type of frog can grow to be 12 inches long. The smallest frog in the world is about 4% as long as the goliath. What is the approximate length of the smallest frog in the world?

    about 0.5 inch

  Holt Mathematics

Practice C
Percent of a Number

Find the percent of each number. Round answers to the nearest tenth, if necessary.

1. 67% of 131

    87.8

2. 19% of 73

    13.9

3. 7% of 129

    9

4. 1.1% of 67

    0.7

5. 128% of 75

    96

6. 227% of 183

    415.4

7. 159% of 271

    430.9

8. 271% of 159

    430.9

9. 17% of 91

    15.5

10. 23% of 77

    17.7

11. 7% of 153

    10.7

12. 7.3% of 73

    5.3

13. 1.95% of 65

    1.3

14. 137% of 81

    111.0

15. 0.03% of 532

    0.2

16. 541% of 132

    714.1

Solve.

17. What number is 55% of 250?

    137.5

18. 39% of 115 is what number?

    44.85

19. 6% of 95 is what number?

    5.7

20. What number is 80% of 860?

    688

21. What number is 41% of 308?

    126.28

22. 125% of 425 is what number?

    531.25

23. 300% of 57 is what number?

    171

24. What number is 180% of 110?

    198

25. In 2004, there were 10,649 commercial radio stations in the United States. About 19.22% of those stations were devoted to country music. About how many country music radio stations were there in 2004?

    about 2,047 radio stations

  Holt Mathematics

  Holt Mathematics

Reteach
Percent of a Number

You can use this proportion to solve percent problems.

$$\frac{part}{total} = \frac{percent}{100}$$

Find 20% of 65.

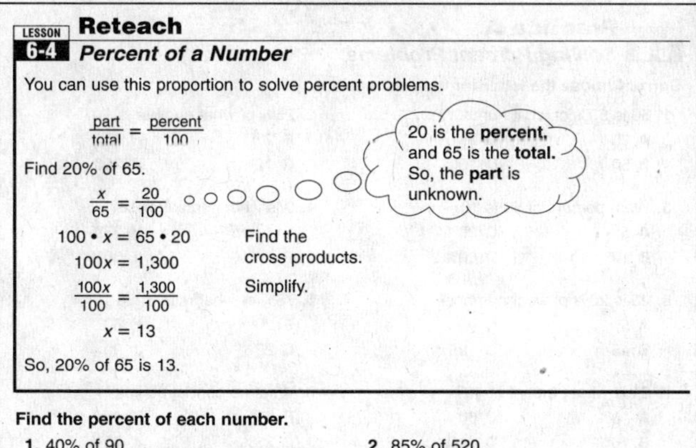

$$\frac{x}{65} = \frac{20}{100}$$

$100 \cdot x = 65 \cdot 20$ Find the cross products.

$100x = 1,300$ Simplify.

$$\frac{100x}{100} = \frac{1,300}{100}$$

$x = 13$

So, 20% of 65 is 13.

Find the percent of each number.

1. 40% of 90

a. part = _____ x

b. total = _____ 90

c. percent = _____ 40

d. $\frac{x}{90} = \frac{40}{100}$

e. $100x =$ _____ 3,600

f. $x =$ _____ 36

2. 85% of 520

a. part = _____ x

b. total = _____ 520

c. percent = _____ 85

d. $\frac{x}{520} = \frac{85}{100}$

e. $100x =$ _____ 44,200

f. $x =$ _____ 442

3. 30% of 80

24

4. 47% of 300

141

5. 45% of 200

90

6. 120% of 70

84

7. 65% of 40

26

8. 25% of 76

19

9. 115% of 40

46

10. 275% of 12

33

31

Holt Mathematics

Challenge
Percent Puzzler

Use the percent clues to find each number below.

1. This number is 8 more than $83\frac{1}{2}$% of 20. _____ 24.7

2. This number is 5 less than 28% of 292. _____ 76.76

3. This number is equal to the sum of 4% of 75 and $32\frac{1}{2}$% of 64. _____ 23.8

4. This number is 7.82 more than $15\frac{3}{4}$% of 320. _____ 58.22

5. This number is equal to the sum of 52% of 86 and 35% of 52. _____ 62.92

6. This number is equal to the product of 8% of 75 and 12% of 60. _____ 43.2

7. This number is equal to the sum of 5% of 125 and $55\frac{1}{2}$% of 20. _____ 17.35

8. This number is equal to the product of 24% of 225 and $3\frac{1}{5}$% of 45. _____ 77.76

9. This number is equal to the sum of $9\frac{3}{10}$% of 110 and $38\frac{1}{4}$% of 124. _____ 57.66

10. This number is 23.04 less than $14\frac{1}{2}$% of 768. _____ 88.32

11. This number is 50.21 more than the sum of $15\frac{1}{2}$% of 42 and 64% of 112. _____ 128.4

12. This number is equal to the difference of 150% of 85 and 280% of 0.3. _____ 126.66

13. This number is equal to the square of $7\frac{1}{2}$% of 160. _____ 144

14. This number is 6.8 less than the product of 5 and 8% of 517. _____ 200

32

Holt Mathematics

Problem Solving
Percent of a Number

Write the correct answer.

The world population is estimated to exceed 9 billion by the year 2050. Use the circle graph to solve Exercises 1–3.

1. What is the estimated population of Africa in the year 2050?

more than 1.8 billion

2. Which continent is estimated to have more than 5.31 billion people by the year 2050?

Asia

3. In the year 2002, the world population was estimated at 6 billion people. Based on research from the World Bank, about 20% lived on less than $1 per day. How many people lived on less than $1 per day?

about 1.2 billion

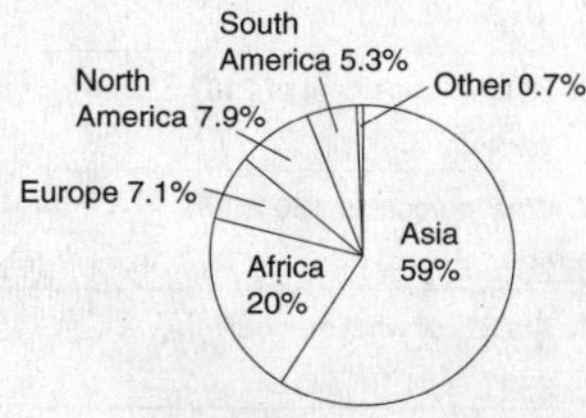

4. What is the combined estimated population for North and South America by the year 2050?

about 1.2 billion

Choose the letter of the best answer.

5. The student population at King Middle School is 52% female. The total student population is 1,225 students. How many boys go to King Middle School?

A 538 boys   C 637 boys
B 588 boys   D 1,173 boys

6. There are 245 students in the seventh grade. If 40% of them ride the bus to school, how many seventh graders do not ride the bus to school?

F 98 students   H 185 students
G 147 students   J 205 students

7. A half-cup of pancake mix has 5% of the total daily allowance of cholesterol. The total daily allowance of cholesterol is 300 mg. How much cholesterol does a half-cup of pancake mix have?

A 100 mg   C 20 mg
B 60 mg   D 15 mg

8. Carey needs $45 to buy her mother a birthday present. She has saved 22% of the amount so far. How much more does she need?

F $35.10   H $23.00
G $44.01   J $9.90

33

Holt Mathematics

Reading Strategies
Find a Pattern

You can use decimals to find the percent of a number.

Find 35% of 80.

35% of 80 means 35% times 80.

Step 1: Change the percent to a decimal by moving the decimal point two places to the left. 35% → 0.35

Step 2: Multiply the decimal times the number. → 0.35 × 80

The answer is 28, so 35% of 80 is 28.

Answer each question.

1. What decimal is equal to 35%? _____ 0.35

2. How do you change a percent to a decimal?

by moving the decimal point two places to the left

3. After the percent is changed to a decimal, what is the next step in finding the percent of a number?

multiplying the decimal times the number

4. Write 10% as a decimal. _____ 0.10

5. What is 10% of 60? _____ 6

6. What is 20% of 60? _____ 12

7. What is 30% of 60? _____ 18

8. What is 40% of 60? _____ 24

9. What pattern did you notice in the answers to 10, 20, 30, and 40 percent of 60?

Possible answer: The answers increased by multiples of 6.

34

Holt Mathematics

Puzzles, Twisters & Teasers
Hi and Dry!

Decide whether or not each equation is correct. Circle the letters above your answers. Then solve the riddle.

1. 7% of 200 = 21

(B)	(Y)
correct	incorrect

2. 16% of 50 = 18

F	(O)
correct	incorrect

3. 8% of 50 = 4

(U)	G
correct	incorrect

4. 5% of 12 = 0.06

K	(R)
correct	incorrect

5. 67% of 90 = 60.3

(S)	L
correct	incorrect

6. 16% of 70 = 1.17

M	(H)
correct	incorrect

7. 20% of 65 = 15

N	(A)
correct	incorrect

8. 0.5% of 36 = 0.18

(D)	P
correct	incorrect

9. 145% of 210 = 304.5

(O)	T
correct	incorrect

10. 1% of 4 = 4

Z	(W)
correct	incorrect

What can fall on the sea without getting wet?

Y O U R

S H A D O W

35
Holt Mathematics

Practice A
Solving Percent Problems

Solve. Choose the letter for the best answer.

1. 50 is 50% of what number?
   A 25   C 75
   B 50   (D) 100

2. 25% of what number is 10?
   F 2.5   (H) 40
   G 15   J 250

3. What percent of 50 is 10?
   A 5%   (C) 20%
   B 10%   D 60%

4. 9 is what percent of 12?
   F 125%   H 33%
   (G) 75% J 30%

5. 25 is 20% of what number?
   A 5   C 95
   B 45   (D) 125

6. 75% of what number is 15?
   (F) 20   H 60
   G 25   J 90

7. 24 is what percent of 40?
   A 10%   (C) 60%
   B 16%   D 64%

8. What percent of 36 is 9?
   F 3%   H 45%
   (G) 25% J 75%

9. What percent of 80 is 48?
   60%

10. What percent of 50 is 5?
   10%

11. What percent of 20 is 16?
   80%

12. What percent of 25 is 15?
   60%

13. 11 is 50% of what number?
   22

14. 40 is 20% of what number?
   200

15. 30 is 25% of what number?
   120

16. 150 is 200% of what number?
   75

17. 60% of what number is 84?
   140

18. 125 is 25% of what number?
   500

19. The sales tax on a $150 DVD player is $6. What is the sales tax rate?
   4%

36
Holt Mathematics

Practice B
Solving Percent Problems

1. 50 is 40% of what number?
   125

2. 12 is 25% of what number?
   48

3. 18 is what percent of 60?
   30%

4. 12 is what percent of 96?
   12.5%

5. 4% of what number is 25?
   625

6. 80% of what number is 160?
   200

7. What percent of 55 is 22?
   40%

8. What percent of 75 is 6?
   8%

9. 15 is 30% of what number?
   50

10. 8% of what number is 2?
   25

11. 7 is what percent of 105?
   $6\frac{2}{3}$%

12. 24 is 40% of what number?
   60

13. 10% of what number is 14?
   140

14. 16 is what percent of 200?
   8%

15. What percent of 32 is 4?
   12.5%

16. What percent of 150 is 60?
   40%

17. 1% of what number is 11?
   1,100

18. 20% of what number is 14?
   70

19. The sales tax on a $750 computer at J & M Computers is $48.75. What is the sales tax rate?
   6.5%

20. A hardcover book sells for $24 at The Bookmart. Ben pays a total of $25.02 for the book. What is the sales tax rate?
   4.25%

37
Holt Mathematics

Practice C
Solving Percent Problems

Solve. Round answers to the nearest tenth, if necessary.

1. 14 is 28% of what number?
   50

2. 28% of what number is 7?
   25

3. 35% of what number is 70?
   200

4. What percent of 63 is 7?
   11.1%

5. 28 is what percent of 210?
   13.3%

6. 65% of what number is 13?
   20

7. What percent of 180 is 36?
   20%

8. 15 is 45% of what number?
   33.3

9. 7 is 2% of what number?
   350

10. What percent of 144 is 108?
   75%

11. What percent of 350 is 840?
   240%

12. 13 is what percent of 77?
   16.9%

13. 11 is 4% of what number?
   275

14. 135% of what number is 115?
   85.2

15. 14 is 16% of what number?
   87.5

16. 12 is 30% of what number?
   40

17. 18 is what percent of 108?
   16.7%

18. 9 is what percent of 39?
   23.1%

19. I am thinking of a number. This number is 18% of 425. What is the number?
   76.5

20. I am thinking of a percent. This percent times 135 gives 40.5 as the result. What is the percent?
   30%

21. Jed buys a sweater that has a sale price of $48. Jed gives the sales clerk $60 and receives $7.80 in change. What is the sales tax rate?
   8.75%

38
Holt Mathematics

Holt Mathematics

Reteach
Solving Percent Problems

You can use this proportion to solve percent problems.

$$\frac{\text{part}}{\text{total}} = \frac{\text{percent}}{100}$$

9 is what percent of 12?

$$\frac{9}{12} = \frac{x}{100}$$

$12 \cdot x = 9 \cdot 100$

$12x = 900$

$\frac{12x}{12} = \frac{900}{12}$

$x = 75$

So, 9 is 75% of 12.

> 9 is the **part**, and 12 is the **total**. So, the **percent** is unknown.

24 is 30% of what number?

$$\frac{24}{x} = \frac{30}{100}$$

$30 \cdot x = 24 \cdot 100$

$30x = 2,400$

$\frac{30x}{30} = \frac{2,400}{30}$

$x = 80$

So, 24 is 30% of 80.

> 24 is the **part**, and 30 is the **percent**. So, the **total** is unknown.

Solve.

1. What percent of 25 is 14?

a. part = __14__

b. total = __25__

c. percent = __x__

d. $\frac{14}{25} = \frac{x}{100}$

e. __25__ $x = $ __1,400__

f. $x = $ __56__

g. 14 is __56%__ of 25.

2. 80% of what number is 16?

a. part = __16__

b. total = __x__

c. percent = __80__

d. $\frac{16}{x} = \frac{80}{100}$

e. __80__ $x = $ __1,600__

f. $x = $ __20__

g. 16 is 80% of __20__.

3. What percent of 20 is 11?

__55%__

4. 18 is 45% of what number?

__40__

5. 15 is what percent of 5?

__300%__

6. 75% of what number is 105?

__140__

Holt Mathematics

Challenge
Percentile Rank

Just as there are three quartiles (the lower quartile, the median, and the upper quartile) that divide a data set into four equal groups, there are 99 *percentiles* that divide a data set into 100 groups.

The definition of a percentile is:

percentile of score $x = \dfrac{\text{number of scores less than or equal to score}}{\text{total number of scores}} \cdot 100$

The frequency table at the right shows the test scores for 28 students.

Find the percentile corresponding to 80.

percentile of 80 $= \dfrac{\text{number of scores less than or equal to 80}}{\text{total number of scores}} \cdot 100$

$= \frac{14}{28} \times 100 = 0.5 \times 100 = 50$

So, 80 is the 50th percentile.

Score	Frequency
100	1
95	2
90	5
85	6
80	7
75	3
70	2
65	2

Use the frequency table to find the percentile corresponding to each score. Round your answer to the nearest whole number.

1. 90

__89__

2. 70

__14__

3. 100

__100__

4. 75

__25__

5. 95

__96__

6. 85

__71__

Use the test scores listed below to find the percentile corresponding to each score. Round your answer to the nearest whole number. (*Hint:* Make a frequency table of the scores.)

84, 77, 77, 77, 92, 77, 84, 84, 95, 84, 68, 92, 84, 100, 77, 77, 84, 92, 77, 92, 92, 95, 77, 68, 84, 100, 92, 84, 95, 92

7. 100

__100__

8. 95

__93__

9. 92

__83__

10. 84

__60__

11. 77

__33__

12. 68

__7__

Holt Mathematics

Problem Solving
Solving Percent Problems

Write the correct answer.

1. At one time during 2001, for every 20 copies of *Harry Potter and the Sorcerer's Stone* that were sold, 13.2 copies of *Harry Potter and the Prisoner of Azkaban* were sold. Express the ratio of copies of *The Prisoner of Azkaban* sold to copies of *The Sorcerer's Stone* sold as a percent.

__66%__

2. A souvenir mug sells for $8.00 at a hotel gift store. Kendra gives the clerk $9.00 and receives $0.56 in change. What is the sales tax rate?

__5.5%__

3. Craig just finished reading 120 pages of his history assignment. If the assignment is 125 pages, what percent has Craig read so far?

__96%__

4. Hal's Sporting Goods had a 1-day sale. The original price of a mountain bike was $325. On sale, it was $276.25. What is the percent reduction for this sale?

__15% off__

Choose the correct letter for the best answer.

5. China's area is about 3.7 million square miles. It is on the continent of Asia, which has an area of about 17.2 million square miles. About what percent of the Asian continent does China cover?

A about 10% C about 17%
B about 15% **D** about 22%

6. After six weeks, the tomato plant that was given extra plant food and water was 26 centimeters tall. The tomato plant that was not given any extra plant food was only 74.5% as tall. How tall was the tomato plant that was not given extra plant food?

F 1.94 cm H 34.89 cm
G 19.37 cm J 48.5 cm

7. In a survey, 46 people, which was 20% of those surveyed, chose red as their favorite color. How many people were surveyed?

A 66 people C 460 people
B 230 people D 920 people

8. Of the 77 billion food and drink cans, bottles, and jars Americans throw away each year, about 65% of them are cans. How many food and drink cans, to the nearest billion, do Americans throw away each year?

F 12 billion **H** 50 billion
G 17 billion J 65 billion

Holt Mathematics

Reading Strategies
Connecting Words and Symbols

You can connect words and symbols to write equations for percent problems.

38	is	20%	of	what number?	← words
38	=	20%	•	n	← symbols
9	is	what percent	of	45?	← words
9	=	n	•	45	← symbols
What percent	of	80	is	10?	← words
n	•	80	=	10	← symbols

Answer each question.

1. What is the symbol for the word "of"?

__•__

2. What symbol means "is"?

__=__

3. What symbol in the above examples stands for the unknown number or percent?

__n__

4. Write the equation for this percent problem using symbols: 6 is 10% of what number?

__$6 = 10\% \cdot n$__

5. Write the words for this equation: $27 = n \cdot 30$.

__27 is what percent of 30?__

6. Write the symbols for this percent problem: What percent of 40 is 30?

__$n \cdot 40 = 30$__

Holt Mathematics

Holt Mathematics

Puzzles, Twisters & Teasers
Rock On!

Fill in the blanks to complete each statement. Then solve the riddle.

L   25 is __25__ % of 100

R   27 is __90__ % of 30

A   105 is __140__ % of 75

O   16 is __33.3__ % of 48

V   56 is 140% of __40__

C   40 is __66.7__ % of 60

I   10 is __12.5__ % of 80

K   9 is __20__ % of 45

T   6 is 10% of __60__

F   7 is 14% of __50__

S   30 is 15% of __200__

E   18 is __300__ % of 6

What kind of concert do pebbles enjoy?

A		R	O	C	K
140		90	33.3	66.7	20

F	E	S	T	I	V	A	L
50	300	200	60	12.5	40	140	25

43
Holt Mathematics

Practice A
Percent of Change

Complete the table.

	Problem	Amount of Change	Original Amount	% of Change
1.	25 is decreased to 17	$25 - 17 = 8$	25	$8 \div 25 = 32\%$
2.	24 is increased to 36	$36 - 24 = 12$	24	$12 \div 24 = 50\%$
3.	50 is decreased to 40	$50 - 40 = 10$	50	$10 \div 50 = 20\%$
4.	40 is increased to 56	$56 - 40 = 16$	40	$16 \div 40 = 40\%$

Find each percent of change. Round answers to the nearest tenth, if necessary.

5. 60 is decreased to 15 __75%__

6. 15 is increased to 21 __40%__

7. 12 is increased to 48 __300%__

8. 100 is decreased to 25 __75%__

9. 60 is decreased to 40 __33.3%__

10. 80 is increased to 152 __90%__

11. 15 is increased to 24 __60%__

12. 72 is decreased to 64 __11.1%__

13. The Big Bike Shop has in-line skates on sale for 15% off the regular price. A pair of in-line skates usually costs $89. Find the amount of the discount and the sale price.

__$13.35; $75.65__

14. In 1935, there were 15,295 banks in the United States. In 2003, there were 9,182 banks. What is the percent of change? Is it a percent increase or decrease?

__40%; percent decrease__

15. A jewelry store is having a going-out-of-business sale. All merchandise is marked 40% off the regular price. The regular price of a watch is $59.95. Find the amount of the discount and the sale price.

__$23.98; $35.97__

16. Last year, there were 381 students at Woodland Middle School. This year, there are 419 students. What is the percent of change? Is it a percent increase or decrease?

__10%; percent increase__

44
Holt Mathematics

Practice B
Percent of Change

Find each percent of change. Round answers to the nearest tenth, if necessary.

1. 20 is decreased to 11 __45%__

2. 24 is increased to 30 __25%__

3. 56 is decreased to 14 __75%__

4. 25 is increased to 100 __300%__

5. 18 is increased to 45 __150%__

6. 90 is decreased to 75 __16.7%__

7. 126 is decreased to 48 __61.9%__

8. 65 is increased to 144 __121.5%__

9. 42 is increased to 72 __71.4%__

10. 84 is decreased to 8 __90.5%__

11. 95 is increased to 145 __52.6%__

12. 248 is decreased to 200 __19.4%__

13. 105 is decreased to 32 __69.5%__

14. 75 is increased to 350 __366.7%__

15. 93 is decreased to 90 __3.2%__

16. 16 is decreased to 2 __87.5%__

17. A backpack that normally sells for $39 is on sale for 33% off. Find the amount of the discount and the sale price.

__$12.87; $26.13__

18. A sporting goods store is having a closeout on a certain style of running shoes. They are marked 55% off the regular price. The regular price is $79.95. Find the amount of the discount and the sale price.

__$43.97; $35.98__

19. A gallery owner purchased a very old painting for $3,000. The painting sells at a 325% increase in price. What is the retail price of the painting?

__$12,750__

20. In August, the Simons' water bill was $48. In September, it was 15% lower. What was the Simons' water bill in September?

__$40.80__

45
Holt Mathematics

Practice C
Percent of Change

Find each percent of change or amount of change. Round answers to the nearest tenth, if necessary.

1. 50 is decreased to 19 __62%__

2. 25 is increased by 80% __20__

3. 40 is decreased to 33 __17.5%__

4. 75 is increased to 200 __166.7%__

5. $80 is increased by 125% __$100__

6. $14.40 is decreased by 20% __$2.88__

7. 11 is decreased by 10% __1.1__

8. 5.7 is increased to 7.8 __36.8%__

9. 42.6 is increased to 71.2 __67.1%__

10. 98 is increased by 85% __83.3__

11. 145 is decreased to 105 __27.6%__

12. 79 is increased to 133 __68.4%__

13. 40 is increased by 12.5% __5__

14. 119 is increased to 277 __132.8%__

15. 537 is decreased to 289 __46.2%__

16. 54 is decreased by 75% __40.5__

17. The school bought 6 new computers for the computer lab. Each computer has a retail price of $979. The school received a discount of 12% on each computer. Find the amount of the discount and the final price for all 6 computers.

__$704.88; $5,169.12__

18. Ali paid $85 for an antique clock several years ago. She recently found that it had increased in value 165%. What is the present value of the clock?

__$225.25__

19. In 1950, there were 40.3 million cars registered in the United States. In 1990, there were 133.7 million cars registered in the United States. What was the percent increase of registered cars from 1950 to 1990?

__231.8%__

46
Holt Mathematics

Holt Mathematics

A change in a quantity is often described as a percent increase or percent decrease. To calculate a percent increase or decrease, use this equation.

$$\text{percent of change} = \frac{\text{amount of increase or decrease}}{\text{original amount}} \cdot 100$$

Find the percent of change from 28 to 42.

- First, find the amount of the change. $42 - 28 = 14$
- What is the original amount?   28
- Use the equation. $\frac{14}{28} \cdot 100 = 50\%$

An increase from 28 to 42 represents a 50% increase.

Find each percent of change.

1. 8 is increased to 22
   amount of change $22 - 8 =$ __14__

   original amount __8__

   $\frac{14}{8} \cdot 100 =$ __175__ %

2. 90 is decreased to 81
   amount of change $90 - 81 =$ __9__

   original amount __90__

   $\frac{9}{90} \cdot 100 =$ __10__ %

3. 125 is increased to 200
   amount of change $200 - 125 =$ __75__

   original amount __125__

   $\frac{75}{125} \cdot 100 =$ __60__ %

4. 400 is decreased to 60
   amount of change $400 - 60 =$ __340__

   original amount __400__

   $\frac{340}{400} \cdot 100 =$ __85__ %

5. 64 is decreased to 48

   __25%__

6. 140 is increased to 273

   __95%__

7. 30 is decreased to 6

   __80%__

8. 15 is increased to 21

   __40%__

9. 7 is increased to 21

   __200%__

10. 320 is decreased to 304

   __5%__

47

Solve.

1. Enrollment in the school soccer league was 340 last year. This year, it dropped 15%. What is the enrollment this year? If enrollment increases 15% next year, what will the enrollment be? Round to the nearest whole number.

   __289; 332__

2. Enrollment in the Wilderness Club increased 8% for each of the last two years. Enrollment two years ago was 175. What was the enrollment for each of the following two years? Round to the nearest whole number.

   __189; 204__

3. The toy store buys stuffed animals from the manufacturer for $2.40 each. The store then sells them at a 96% increase in price. What is the retail price of each animal? If the customer brings in a coupon for 20% off the retail price, how much will the stuffed animal cost?

   __$4.70; $3.76__

4. Enrollment in the Jacksonville Township Basketball League was 425 last year. This year, the number enrolled increased 12%. If enrollment drops 12% next year, how many people will be enrolled in the league? Round to the nearest whole number.

   __419__

5. The manager of a card store buys cards from the manufacturer for $0.65 each. The store regularly sells the cards at a markup of 130%. For a sale, the manager marks down half of the cards 15%. What is the price of the cards that are on sale?

   __$1.27__

6. Jamal wants to buy a video game. Two stores in the neighborhood offer the game at $49.95. Jamal has a coupon for 18% off the regular price at Fun Electronics. The other store, Gamers, is having a sale with 8% off everything in the store. Jamal also has a coupon for 10% off the sale price at Gamers. At which store should he buy the game? Explain.

   __Fun Electronics, since the game will cost $40.96;__

   __it will cost $41.36 at Gamers.__

48

Write the correct answer.

1. In 2002, U.S. consumers bought about 8.1 million new cars. In 2003, that number decreased by about 6%. To the nearest hundred thousand, or tenth of a million, how many new cars did U.S. consumers buy in 2003?

   __about 7.6 million cars__

2. In Union County, Florida, the 1990 census listed the population at 10,252. The 2000 census listed the population as 13,442. What percent increase is this to the nearest tenth of a percent?

   __31.1% increase__

3. World production of motor vehicles increased from about 60 million in 2002 to 62 million in 2003. What was the percent increase to the nearest percent?

   __3% increase__

4. Arthur's dog, Shep, used to weigh 158 pounds. The vet put him on a diet and he lost 13% of his weight. To the nearest pound, how much does Shep weigh now?

   __137 pounds__

5. The number of volunteers rose from 47 on Monday to 64 on Tuesday. What is the percent increase to the nearest tenth of a percent?

   __36.2% increase__

6. Coretta's bowling average decreased from 158 to 133. What is the percent decrease to the nearest tenth of a percent?

   __15.8% decrease__

Choose the correct letter for the best answer.

7. Shandra scored 75 on her first math test. She scored 20% higher on her next math test. What did she score on the second test?
   A 80
   B 85
   Ⓒ 90
   D 95

8. Jim's Gym had income of $20,350 last month. The total increased by $2,000 this month. What was the percent increase to the nearest percent?
   F 11%
   Ⓖ 10%
   H 7%
   J 6%

9. Last year, the average number of absences in school was 8 students per day. This year, the absentee rate is down to 6 students per day. What is the percent decrease in student absences this year?
   A 75%
   B 33%
   Ⓒ 25%
   D 66%

10. During the 2002-2003 ski season $171 million worth of snowboarding equipment was sold. Sales increased by about 15% during the 2003-2004 season. About how much were sales of snowboarding equipment in the 2003-2004 season?
   F $145 million
   G $156 million
   H $186 million
   Ⓙ $197 million

49

Percent can be used to describe change. It is shown as a ratio.

$$\text{Percent of change} = \frac{\text{amount of change}}{\text{original amount}}$$

The following steps describe how the percent of change is figured on a savings account that starts with $50.

This is the original amount in the account: $50
This is the current amount in the account: $30
This is the amount that the account decreased by: $20

$\text{Percent of change} = \frac{\$20}{\$50}$ the amount of change over the original amount

Savings went down, so this ratio is → the **percent of decrease** in savings.

1. How much money was placed into the savings account when it opened?

   __$50__

2. Did the number of dollars in the account increase or decrease?

   __decrease__

3. When you are figuring the percent of change, where is the original number placed in the fraction?

   __in the denominator (or bottom part) of the fraction__

Use this information for Exercises 4–6: A clothing salesman sold 25 shirts his first day on the job and 45 shirts the second day.

4. What is the original number of shirts he sold?

   __25__

5. How many more shirts did he sell the second day than the first day?

   __20__

6. Write the fraction that shows the amount of change over the original amount.

   __$\frac{20}{25}$__

50

Puzzles, Twisters & Teasers

Be Patient!

Circle words from the list in the word search. Then find a word that answers the riddle. Circle it and write it on the line.

percent	change	increase	decrease	round
tenth	substitute	decimal	discount	

When do dentists get angry?

When they run out of P A T I E N T S

51   **Holt Mathematics**

Practice A

Simple Interest

Find each missing value.

1. $p = \$1{,}000$, $r = 5\%$, $t = 2$ years

$I = $ **$\$1{,}000$** $\cdot$ **0.05** $\cdot$ **2**

$I = $ **$\$100$**

2. $p = \$600$, $r = 4\%$, $t = 3$ years

$I = $ **$\$600$** $\cdot$ **0.04** $\cdot$ **3**

$I = $ **$\$72$**

3. $I = \$330$, $r = 3\%$, $t = 1$ year

$\$330$ $= p \cdot$ **0.03** $\cdot$ **1**

$p = $ **$\$11{,}000$**

4. $I = \$270$, $r = 5\%$, $t = 3$ years

$\$270$ $= p \cdot$ **0.05** $\cdot$ **3**

$p = $ **$\$1{,}800$**

5. $I = \$600$, $p = \$2{,}500$, $t = 4$ years

$\$600$ $= \$2{,}500 \cdot r \cdot$ **4**

$r = $ **6%**

6. $I = \$108$, $p = \$900$, $t = 3$ years

$\$108$ $= \$900 \cdot r \cdot$ **3**

$r = $ **4%**

7. $p = \$250$, $r = 6\%$, $t = 5$ years

$I = $ **$\$75$**

8. $p = \$3{,}000$, $r = 7\%$, $t = 4$ years

$I = $ **$\$840$**

9. $I = \$750$, $r = 4\%$, $t = 5$ years

$p = $ **$\$3{,}750$**

10. $I = \$696$, $r = 3\%$, $t = 4$ years

$p = $ **$\$5{,}800$**

11. $I = \$425$, $p = \$1{,}700$, $t = 5$ years

$r = $ **5%**

12. $I = \$1{,}680$, $p = \$12{,}000$, $t = 2$ years

$r = $ **7%**

13. You deposit \$5,000 in an account that earns 5% simple interest. How long will it be before the total amount is \$6,000?   **4 years**

14. After 6 years, an account that earns 4% simple interest has earned \$480 in interest. How much was the initial deposit?   **\$2,000**

15. A deposit of \$7,500 earns \$3,900 over a period of 8 years. What is the simple interest rate?   **6.5%**

16. You deposit \$4,500 in an account that earns 6% simple interest. How much will be in your account after 5 years?   **\$5,850**

52   **Holt Mathematics**

Practice B

Simple Interest

Find each missing value.

1. $p = \$1{,}500$, $r = 5\%$, $t = 3$ years

$I = $ **$\$225$**

2. $p = \$6{,}000$, $r = 4\%$, $t = 2$ years

$I = $ **$\$480$**

3. $I = \$30$, $r = 4\%$, $t = 2$ years

$p = $ **$\$375$**

4. $I = \$180$, $r = 5\%$, $t = 3$ years

$p = $ **$\$1{,}200$**

5. $I = \$20$, $p = \$250$, $t = 2$ years

$r = $ **4%**

6. $I = \$144$, $p = \$800$, $t = 3$ years

$r = $ **6%**

7. $p = \$525$, $r = 3\%$, $t = 1$ year

$I = $ **$\$15.75$**

8. $p = \$3{,}200$, $r = 6\%$, $t = 4$ years

$I = $ **$\$768$**

9. $I = \$450$, $r = 6\%$, $t = 4$ years

$p = $ **$\$1{,}875$**

10. $I = \$1{,}440$, $r = 3\%$, $t = 5$ years

$p = $ **$\$9{,}600$**

11. $I = \$1{,}275$, $p = \$5{,}100$, $t = 5$ years

$r = $ **5%**

12. $I = \$3{,}920$, $p = \$14{,}000$, $t = 4$ years

$r = $ **7%**

13. $p = \$1{,}300$, $r = 4.5\%$, $t = 6$ months

$I = $ **$\$29.25$**

14. $I = \$47.25$, $r = 3.5\%$, $t = 1.5$ years

$p = $ **$\$900$**

15. $I = \$891$, $p = \$2{,}700$, $t = 5.5$ years

$r = $ **6%**

16. $I = \$126$, $p = \$400$, $t = 9$ years

$r = $ **3.5%**

17. You deposit \$2,500 in an account that earns 4% simple interest. How long will it be before the total amount is \$3,000?   **5 years**

18. You deposit \$5,000 in account that earns 6.5% simple interest. How much will be in the account after 3 years?   **\$5,975**

19. A deposit of \$10,000 was made to an account the year you were born. After 12 years, the account is worth \$16,600. What simple interest rate did the account earn?   **5.5%**

20. How long will it take for \$6,500 to double at a simple interest rate of 7%? Round to the nearest tenth of a year.   **14.3 years**

53 .   **Holt Mathematics**

Practice C

Simple Interest

Complete the table.

	Principal	Interest Rate	Time	Simple Interest
1.	\$2,000	3.5%	3 years	**$210**
2.	\$1,250	**7%**	4 years	\$350
3.	\$500	4.5%	**7 years**	\$157.50
4.	**$9,400**	5%	54 months	\$2,115
5.	\$12,000	**4.5%**	2 years	\$1,080
6.	\$1,800	7.5%	6 months	**$67.50**
7.	**$307.50**	6%	4 years	\$73.80
8.	\$8,500	6.5%	**12 years**	\$6,630
9.	**$4,000**	3.5%	5 years	\$700
10.	\$3,300	4.75%	**2 years**	\$313.50
11.	\$6,800	**2.5%**	16 years	\$2,720
12.	\$2,400	5.5%	30 months	**$330**

Solve.

13. A deposit of \$500 in an account earns 6% simple interest. How long will it be before the total amount is \$575?   **2.5 years**

14. What simple interest rate is needed for \$1,000 to grow to \$1,071.25 in 9 months?   **9.5%**

15. A deposit of \$3,000 becomes \$3,810 after 6 years. What is the simple interest rate on the account?   **4.5%**

16. How long will it take for \$1,000 to double at a simple interest rate of 5.5%?

about 18 years

17. Consuelo deposited an amount of money in a savings account that earned 6.3% simple interest. After 20 years, she had earned \$5,922 in interest. What was her initial deposit?   **\$4,700**

18. A deposit of \$2,500 grew to \$3,325 after 6 years. What is the final value of a deposit of \$7,500 at the same interest rate for the same period of time?   **\$9,975**

54   **Holt Mathematics**

Holt Mathematics

Reteach
Simple Interest

When you put money into a bank account, you may receive simple interest for loaning the bank your money.

$$\text{Interest} = \text{Principal} \cdot \text{Rate} \cdot \text{Time}$$
$$I = p \cdot r \cdot t$$

You can use the expression $\dfrac{I}{p \cdot r \cdot t}$ to solve interest problems.

- To find interest (I), put your finger over I. Perform the operations for letters you see.
- To find principal (p), put your finger over p. Perform the operations for letters you see.
- To find interest rate (r), put your finger over r. Perform the operations for letters you see.

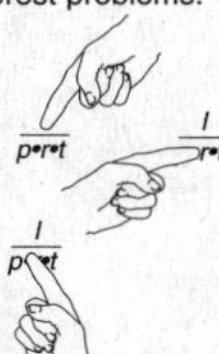

Find each missing value.

1. $p = \$400$, $r = 5\%$, $t = 3$ years

$I = p \cdot r \cdot t$

$I = \underline{400} \cdot \underline{0.05} \cdot \underline{3}$

$I = \underline{\$60}$

2. $p = \$15,000$, $r = 6\%$, $t = 2$ years

$I = p \cdot r \cdot t$

$I = \underline{\$15,000} \cdot \underline{0.06} \cdot \underline{2}$

$I = \underline{\$1,800}$

3. $I = \$350$, $r = 7\%$, $t = 2$ years

$p = \dfrac{I}{r \cdot t}$

$p = \dfrac{350}{0.07 \cdot 2}$

$p = \underline{\$2,500}$

4. $I = \$168$, $p = \$1,400$, $t = 4$ years

$r = \dfrac{I}{p \cdot t}$

$r = \dfrac{168}{1,400 \cdot 4}$

$r = \underline{0.03} = \underline{3\%}$

5. $I = \$57$, $p = \$380$, $t = 5$ years

$r = \underline{3\%}$

6. $p = \$4,800$, $r = 6\%$, $t = 2$ years

$I = \underline{\$576}$

7. $I = \$1,200$, $r = 4\%$, $t = 4$ years

$p = \underline{\$7,500}$

8. $p = \$750$, $r = 7\%$, $t = 3$ years

$I = \underline{\$157.50}$

Holt Mathematics

Challenge
Adding On

Simple interest is the amount of interest earned on the original principal. However, you can earn interest on the interest as well as on the principal. This is called **compound interest**.

Josef deposits $400 in a bank that pays 5% interest, compounded annually. Find the amount of interest and principal in Josef's account after 3 years.

Interest and Principal

Year 1	Year 2	Year 3
$I = p \cdot r \cdot t$	$I = p \cdot r \cdot t$	$I = p \cdot r \cdot t$
$= 400 \cdot 0.05 \cdot 1$	$= 420 \cdot 0.05 \cdot 1$	$= 441 \cdot 0.05 \cdot 1$
$= 20$	$= 21$	$= 22.05$
Interest is $20.	Interest is $21.	Interest is $22.05.
Principal is $420.	Principal is $441.	Principal is $463.05.

The amount of interest earned after 3 years is $20 + $21 + $22.05 = $63.05.

The amount of principal plus interest after 3 years is $400 + $63.05 = $463.05.

You can also find the total of principal and interest on $400 for 3 years at 5% by multiplying on a calculator 400 • 1.05 • 1.05 • 1.05. This shows the amount of interest and principal in Josef's bank account after 3 years, or $463.05.

Find the total amount of interest and principal if interest is compounded annually.

1. $500 for 2 years at 4%

$\underline{\$540.80}$

2. $1,200 for 3 years at 4.5%

$\underline{\$1,369.40}$

3. $300 for 4 years at 5.5%

$\underline{\$371.65}$

4. $750 for 2 years at 5.5%

$\underline{\$834.77}$

5. $98 for 3 years at 6.5%

$\underline{\$118.38}$

6. $1,056 for 4 years at 5.25%

$\underline{\$1,295.84}$

7. $520 for 5 years at 4.75%

$\underline{\$655.80}$

8. $873 for 6 years at 5.2%

$\underline{\$1,183.34}$

Holt Mathematics

Problem Solving
Simple Interest

Write the correct answer.

Use the graph to solve Exercises 1–3.

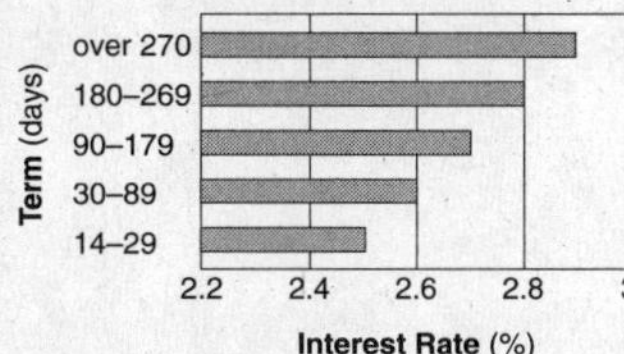

AEA Bank Simple Interest CDs

Term (days): over 270, 180–269, 90–179, 30–89, 14–29

Interest Rate (%): 2.2, 2.4, 2.6, 2.8, 3

1. How much more interest would be earned on a $100,000 CD for 9 months than for 6 months?

$\underline{\$775}$

2. A customer earned $3,262.50 interest on a 9-month CD. How much was the opening deposit?

$\underline{\$150,000 \text{ deposit}}$

3. Mrs. Wallace bought a $125,000 CD with a term of 3 years. How much will she earn in 3 years?

$\underline{\$10,875}$

4. Until June 2002, the simple interest rate on Stafford loans to college students was 5.39% while the student was still in college. How much interest would a student pay on a $1,500 loan for 2 years?

$\underline{\$161.70}$

5. Diego deposits $4,200 into a savings account that pays 5% simple interest. He decides not to touch the money until it doubles. How long will Diego have to keep the money in this account?

$\underline{20 \text{ years}}$

Choose the letter of the best answer.

6. Scott took out a 4-year car loan for $5,500. He paid back a total of $7,370. What interest rate did he pay for this loan?

A 9.5%   **C** 8.5%
B 9%   D 7.5%

7. How much interest would you earn if you were to deposit $575 for 3 months at 2.88% simple interest?

F $4.14   H $41.40
G $4.83   J $48.30

8. How long would you need to keep $775 in an account that pays 3% simple interest to earn $93 interest?

A 4 years   C 4 months
B 2 years   D 2 months

9. If you borrow $12,000 for 30 months at 6.5% simple interest, what is the total amount you will have to repay?

F $12,065   **H** $13,950
G $12,780   J $21,500

Holt Mathematics

Reading Strategies
Focus on Vocabulary

Principal is the amount of money you save or borrow from a bank.

Interest is the amount of money the bank pays you for the use of your money, or the amount of money you pay the bank to borrow its money.

Rate is the amount of interest paid on money you save or borrow. Rates are usually given as percents.

For Exercises 1–3, write principal, interest, or rate to identify each situation.

1. You have $250 in a savings account.

$\underline{\text{principal}}$

2. The bank pays you 4% a year on the money you have saved.

$\underline{\text{rate}}$

3. The bank paid you $10 on your savings account last year.

$\underline{\text{interest}}$

If you wanted to borrow $3,000 for two years at a rate of 6%, you could use this formula to find the amount of interest you would pay:

$$\text{Interest} = \text{principal} \cdot \text{rate} \cdot \text{time}$$
$$I = p \cdot r \cdot t$$
$$I = \$3,000 \cdot 6\% \cdot 2$$
$$I = \$3,000 \cdot 0.06 \cdot 2 \quad \leftarrow \text{Change percent to decimal.}$$
$$I = \$360 \quad \leftarrow \text{Multiply.}$$

4. What formula is used to find the interest for this loan?

$\underline{I = p \cdot r \cdot t}$

5. What is the decimal for 6%?

$\underline{0.06}$

6. How do you find the amount of interest that will be paid on a loan?

$\underline{\text{Multiply the principal by the rate by the length of time on the loan.}}$

Holt Mathematics

Puzzles, Twisters & Teasers

Simply Interesting!

Use what you know about simple interest to complete the chart.
Then use your answers and the answer key to solve the riddle.

Principal	Interest Rate	Time	Simple Interest
$3,455	4%	**5 years** S	$691
$6,000 F	5.25%	2 years	$630
$16,500	**7%** E	24 months	$2,310
$750 A	6%	3 years	$135
$625	3.5%	10 years	**$218.75** C

What do people in clock factories do all day?

They make ___F___ ___A___ ___C___ ___E___ ___S___ .
      $6,000 $750 $218.75 7% 5 years

59

Holt Mathematics